EXPLORING MATHEMATICS WITH MICROCOMPUTERS

MEP Readers 8

EXPLORING MATHEMATICS
WITH MICROCOMPUTERS

edited and introduced by
Nigel Bufton

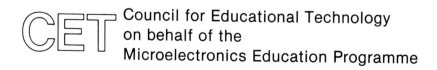

Council for Educational Technology
on behalf of the
Microelectronics Education Programme

Published and distributed by the Council for Educational Technology,
3 Devonshire Street, London W1N 2BA on behalf of the
Microelectronics Education Programme, Cheviot House, Coach Lane
Campus, Newcastle upon Tyne NE7 7XA

First published 1986

© Council for Educational Technology 1986

ISSN 0264-4142
ISBN 0 86184-162-X

 British Library Cataloguing in Publication Data

Exploring mathematics with microcomputers —
  (MEP Readers; ISSN 0264-4142; v.8)
  1. Mathematics — Data processing —
  Study and teaching
  2. Microcomputers — Study and teaching
  I. Bufton, Nigel   II. Council for
  Educational Technology for the United
  Kingdom   III. Microelectronics Education
  Programme   IV Series
  510'.28'5416   QA76.95
  ISBN 0-86184-162-X

Printed in Great Britain by
H Cave & Co Ltd
Cumberland Street
Leicester LE1 4QQ

Contents

List of figures

Preface

'Mathematics is a difficult subject both to teach and to learn.'
Cockcroft (para 228).

Those who have been involved in the teaching of mathematics are likely to confirm the truth of this statement. However, despite it or even because of it, the majority of us still remain active and involved teachers. If we are sustained through the pleasures we derive from mathematics and from working at it with children then the microcomputer offers us a resource with which we can enhance these pleasures. The relationship between mathematics and the micro in schools is already well established and an ever-growing array of software and related material is available to us to support our teaching.

This book is one of a series of Readers commissioned by the Microelectronics Education Programme to provide information and ideas on the use of micros in the classroom. It aims to offer the reader a look at some of the ways that the micro has been, is being and can be used in the mathematics classroom, and considers the effect this has on the teaching and learning of mathematics. It is not intended to provide any dogmatic statements or prescriptive accounts on how to use a micro nor to discuss the relative merits of particular hardware or programming languages. The contributors have written about their experiences and beliefs and in doing so invite readers to go and try for themselves.

The general philosophy that runs through this book when consideration is given to the use of the micro in the teaching of mathematics might best be summed up by the Chinese proverb 'Tell me; I forget.
Show me; I remember.
Involve me; I understand.'

Acknowledgements
Thanks go to all those who have contributed to this book; to Mary Brading for her cartoon; to Richard Fothergill for providing me with the opportunity to produce it and for the patience of colleagues and family.

Nigel Bufton
Mathematics Curriculum Development Centre
West Sussex Institute of Higher Education
October 1985

Introduction

Nigel Bufton, West Sussex Institute of Higher Education

It is sometimes easy to forget that it is only during this decade that the microcomputer has become a common resource in many of our classrooms. In 1980, only five per cent or so of secondary schools had access to a computer; in primary schools they had made no significant impact. In 1981, the situation improved dramatically, in the main through the financial support of the Department of Industry to our secondary schools. During that year, the Department of Education and Science established the Microelectronics Education Programme, MEP, to help teachers use this new technology in their classrooms to aid and encourage children in their learning, and to support schools as they prepared their pupils to face the new technological society, where microelectronics would be a significant feature in the control of our daily lives. MEP quickly began producing a range of in-service material which included appropriate software, and its Microprimer packs were distributed free to those schools participating in the DoI scheme. In 1982, greater attention was directed towards the primary schools, and the DoI's financial support was extended to include them. In 1983 the MEP National Primary Project was established, its aim being to identify the needs of primary-school pupils and teachers and see how available technology might be used to add to the learning experiences of children. Since its formation it has produced a wide range of material for primary schools including some useful mathematical software, and organized a number of in-service courses for teachers. Towards the end of 1984, nearly all schools, primary and secondary, owned their own computer. It is ironic perhaps that this was the very year used by Orwell for the title of his book in which he warns us about 'the danger inherent in the machine' leading us to the situation where everybody would 'become literate and learn to think for themselves'!

At the same time as this technology was being introduced into the classroom, a major appraisal was being conducted into the teaching of mathematics. This was to result in the Report of the Committee of Inquiry into the Teaching of Mathematics in Schools — the Cockcroft report — published in 1982. It was to pay particular regard to the mathematics needed in employment and during our adult life. For many, mathematics induces feelings of anxiety and guilt and, when asked to undertake even the simplest of tasks, many appear helpless and exhibit a low level of competency in the subject. The report observes that for them, mathematics teaching had been unsuccessful and even counter-productive. It offers a number of recommendations for change in an attempt to begin to remedy this situation.

Unfortunately, these two major events, the emergence of the micro in the classroom and the Cockcroft inquiry, were too coincidental for there to be much information available to the Committee on the use of micros in the teaching of mathematics. The report recognized that micros would offer 'considerable opportunity to teachers of mathematics both to enhance their existing practice and also to work in ways which have not hitherto been possible' (para 402), but points out that the possession of a micro 'will not itself improve the teaching of mathematics' (para 404). The report does, however, offer much sound advice within the short section devoted to computers.

1983 saw Dr T J Fletcher, HMI, produce a discussion paper on 'Microcomputers and mathematics in schools'. The paper recognized that there had been a rapid development in the use of micros in the classroom. It offered a description of some of the practices that were found in a number of visited schools and commented on the influence of the micro in changing classroom approaches to the methods of teaching mathematics. The author observed that the greatest change was to be seen in the 'response of the pupil' and that pupils 'in surprisingly large numbers were finding a joy and zest in some aspects of mathematics which they did not find before' (para 3). The paper suggested that the greatest challenge open to teachers of mathematics was to develop this enthusiasm in others and maintain it, thereby generating a 'direction and continuity from these spontaneous beginnings', adding that it was usually dampened when the activity was 'institutionalized in the wrong way and the pupils [were] no longer making decisions themselves' (para 110).

In May 1983, a Curriculum Conference was held at Pendley Manor, Hertfordshire, jointly sponsored by the DES and MEP. There a group of people involved in mathematics and education looked closely at the relationship between mathematics and the micro in the classroom. Their report, known as the Pendley Manor Report, was to offer advice to MEP on future policy, and has appeared in the Mathematical Association's 'Mathematics in schools'. It points out how strong links have already been forged between mathematics and micros through the way that mathematics relates to the operation and working of the machine and by pupils gaining greater access to mathematics through writing programs of their own. One of the recommendations made was that 'specific attention should be given to programs which promote those classroom activities which the Cockcroft report (para 243) has identified as missing from the majority of classrooms, namely: discussion, problem-solving and investigational activities'. It also adds that the teaching of narrow, well established skills and techniques traditionally practised in our mathematics classrooms is inappropriate to the needs of a technological society; rather, that children should be given an opportunity to make a general study of 'algorithms and procedures' through exploring, modifying and designing them.

More recently, in 1985, *Mathematics from 5 to 16* was published. This HMI discussion document acknowledges the importance of the role of the micro in the teaching of mathematics and states that the school mathematics 'content should be influenced increasingly by developments in microcomputing', as they provide 'powerful means of doing mathematics extremely quickly and in a visually dramatic way' (p34). In its review of appropriate classroom resources micros are considered to be 'essential' (p43), and within the overall framework of objectives for the teaching of mathematics it includes the 'use of microcomputers in mathematical activities' (p13).

Before the 1980s a decision had been taken by the Government of the day to introduce a single system of examination at 16 plus. In 1978, the Government paper, *Secondary School Examinations: a single system at 16 plus,* stated that the proposed 'examination system should enable all candidates to demonstrate their capabilities'. The Cockcroft report outlines many of the limitations the old system had through its total reliance upon timed written examinations — in particular their inability to assess a candidate's 'ability to undertake practical and investigational work or ability to carry out work of an extended nature' (para 532). Since, it adds, these are best assessed in the classroom, so 'provision should be made for an element of teacher assessment to be included in the examination of pupils of all levels of attainment' (para 535). The Secondary Examinations Council in its submission to the Secretary of State in 1983 welcomed the call 'for an increased role for course work and school- or college-based assessment of various kinds'. From this lengthy debate emerged the GCSE and subsequently the 'national criteria' to be used for the assessment of mathematics. Sadly and neglectfully, they fail to make any direct reference to the micro. Included, however, is school-based assessment and the requirement that all candidates must carry out some practical and investigational work and undertake and sustain work of an extended nature. These will become compulsory in 1991, but clearly there is a need to begin this process as soon as possible in readiness for these changes. The micro has a valuable contribution to make to many of these innovations.

In summary, over this short period of time we have been witness to:

— the continuing development of cheaper, compact yet more powerful micros with much improved graphics capabilities

— an increase in their availability within the home and school to the extent that every school now has access to at least one machine

— a recognition that a micro is an essential item for resourcing mathematics teaching and that there already exists a strong relationship between the two

— an acceptance that in the past mathematics teaching had been unsuccessful for many people and that changes must be made in classroom practices if this situation is to improve

— the emergence of the micro as an important influence for, and strong initiator to, effecting change in the teaching of mathematics

— the establishing of a new assessment system within which the micro can fulfil a major role.

This book can only provide a sample of what is happening in relation to this constantly shifting backcloth. There is a growing collection of valuable examples which highlight how a micro can be used in the classroom to stimulate, encourage and assist pupils in their mathematical ventures. Professional associations now focus upon the use of the micro within their publications. These publications are referred to throughout the book and offer a source of material and ideas, many of which have been developed from within the classroom. It is not the intention of this book to make the reader think that all mathematics must, or indeed should, be taught using a micro for that is an unacceptably narrow view which fails to acknowledge the value of other learning and teaching resources. Rather, in recognizing that the micro offers a unique freedom for mathematical exploration, it would claim that all pupils are entitled to share in this experience.

Software

Much of the software that was available at the beginning of the decade was of doubtful educational value, and quickly became dated. Software houses may come and go but software continues to be produced, presenting the prospective purchaser with a formidable array of material from which to choose. The biggest problem about purchasing any software is in deciding whether it is worthwhile before making the investment. For a small primary school this can be an important and costly decision. Fortunately a variety of magazines now contain reliable reviews and detailed guidance on software, but word-of-mouth and local reputation hold great sway. MEP has helped through establishing its regional centres where teachers can review and evaluate available resources, and MAPE (Micros and Primary Education) offers advice and support through its regional representatives. A number of mathematics schemes currently available at both primary and secondary level also include software packages within them. Usually these have been carefully constructed to fit into the overall teaching model suggested and are intended to enhance or support it.

A view commonly expressed is that many of the claims made by the producers of software are not met when the material is used in classrooms. A particular piece of software might well be designed to offer a wide range of challenges or be flexible enough to be applied to a variety of situations, where and when the teaching approach and classroom management facilitate its appropriateness and accessibility. However, these teaching styles and classroom experiences are not as readily transferable as the cassette or disc, and the full potential of the resource may not be reached. A big chunk of marketed software still claims to help in the teaching of addition, manipulation of fractions or some such skill through providing an endless variety of simple display/respond/assess/correct-type questions supported by a dazzle of colour and shrill musical interludes. These blatant skill-and-drill programs offer practice and might strengthen techniques but we already have at our disposal worksheets and textbooks which we use to do the same thing. Those programs that set such manipulative skills in the context of a game, a simulation, or a problem, have relegated them to second place, putting the formulation and deployment of strategies and problem-solving skills as their primary goals and thereby raising the quality of thinking. They do offer practice but make use of an expensive and sophisticated resource to enable children to experiment with mathematical ideas and provide them with a medium where they can make mistakes and learn from them.

There can be no definitive way to use a piece of software. In some classrooms, teacher and pupils are able to take software which would appear to many to be narrow and limited and by providing a suitable context for it and posing appropriate questions turn it into a very stimulating and flexible resource. Observing young children use a program which had neither a clear list of instructions nor any support material to explain its possibilities soon demonstrated how creative they were and how ready they were to turn the mundane into their own exciting tool. The program was an incomplete part of a package to help the user become more familiar with the arithmetic of the circle. It asked for three numbers to be input and then drew a coloured disc with the circumference and diameter highlighted in different colours. We later learned that the user would then be invited to input values for the diameter and the circumference which would be checked and responded to. However, this bit was missing. The program soon became a testbed for the children who spent a long time setting themselves tasks involving them in the designing of circular patterns. They did learn a lot about number, coordinates, order, etc, but we did not record the extent to which they improved their abilities to calculate circumferences.

Reference is made to the use of appropriate software in a number of chapters in this book. Richard Phillips considers how a single micro may be used and raises issues concerning classroom management. He selects examples of software he has used and examines the reasons these were successful. James Whitbread had one micro between all the pupils in a primary school. He presents his experiences with LOGO and discusses how limited resources can provide every pupil with an opportunity for hands-on experience.

Graphics
The quality of the graphics capabilities of micros is improving with each new machine produced. The speed of drawing, the colours available and the quality of resolution are features that are emphasized by the manufacturers. The graphics facilities do raise the status of the micro. Gone is the old image of the computer as just a fast and furious number cruncher. In its more elevated position it provides the dynamic picture show with words and music. Mathematics exploits this facility in a variety of ways. In attempting to harness the graphics and use the commands available it is necessary to become familiar with the mathematics of the machine. How does it map its screen? How can I move the origin? What colour combinations are available? Etc. All this has an inherent mathematical structure which has to be understood. The next stage is applying this to mathematical problems. How can I draw a square, hexagon, circle? How can I transform my circle into an ellipse? How can I rotate my ellipse? This requires some understanding of the mathematics of the ellipse. There also has to be an ability to translate this knowledge into a form that can be used on the micro. Software is available that carries out this translation for us. Families of curves can be investigated using graph plotters where the explicit form of the function can be input and the graph drawn, their discontinuities coped with too. Implicit forms may still require our intervention in the process. Data can be input and graphical representations produced.

The turtle graphics of LOGO offers a different inherent mathematical form. This makes the task of translation more direct in some cases — for example, when drawing the regular polygons — but the form also influences the background mathematical understanding that is required before the translation takes place. What is the mathematical understanding we must have if we are to draw a circle with some given radius, and transform this into an ellipse using turtle graphics? LOGO offers the user an opportunity to program recursively and young children quite happily make use of this when they construct their graphics procedures. This is usually employed as a means of creating a cycle of operations, as commands for changing the order of program execution do not exist as in BASIC. There is a suggestion that children can be taught to think recursively within LOGO but there is hot debate as to whether this skill is transferable or even applicable to other mathematical contexts.

In his article, Trevor Fletcher considers the way that mathematical graphics can be introduced into the classroom using a do-it-yourself approach. With a number of short programs to control graphics output, he offers a context for the learning of mathematics and the meeting of new ideas. These include curve stitching; the graphing of implicit and explicit functions; transformation geometry; Lissajou and other figures and the use of moving pictures to display coupled oscillation.

Seamus Dunn takes a detailed look at the mental imagery used in the teaching and learning of mathematics. The micro offers pictorial imagery and Seamus seeks to establish the truth or otherwise of the view that difficult ideas in mathematics can be made accessible to children if they are presented pictorially. He relates his experiences with groups of nine-year-old children working on a micro with programs that display a rotating arm and a sine wave representing its vertical displacement. The results of this pilot study offer many useful guidelines for further study. He makes reference to Seymour Papert who has written about the way children use the 'microworlds' created, to develop an 'intuitive understanding' of ideas from the pictorial imagery these offer.

Norman Blackett and David Tall look at sixth-form mathematics and examine how the micro can improve geometric skills and understanding. They use a software package to promote a greater appreciation of curves and graphs. From this established base they show how the introduction of calculus to students becomes a natural extension of curve sketching and how a topic that often causes many difficulties in understanding can be demystified. They further consider the implications the micro has on other areas of the A-level syllabus.

Algebra

In its answer to the question 'Why teach mathematics?' the Cockcroft report concludes that the principal reason is that mathematics 'provides a means of communication which is powerful, concise and unambiguous' (para 3). This power is as much an appeal for some as it is an inhibitor for others. The conciseness of symbols and their systematic use allow the mathematician to work in an environment where the abstract and general can be expressed and developed. For children this is very often the greatest hurdle as the symbolizing is introduced before the need for it is created. Children first speak when they learn that it will result in some action, and they see that that is the best way to get it. The micro provides this kind of encouragement for symbolizing: in order to get the necessary action we must use recognized symbolic instruction. Home computers are still sold and children are their main users. The child who has access to one is often readily motivated towards mathematics as micros can act as favourable influencers of attitude. The child who leaves it on overnight to find the millionth triangle number has already begun to symbolize and will be appreciative of the power of conciseness. The language of programming provides us with the environment where children can see the need for working with general statements and for avoiding ambiguity: now algebra may be introduced for a purpose.

John Higgo discusses how he naturally turns to a micro when teaching algebra and offers the reader examples of how this has helped his pupils gain greater understanding and cope with the manipulation of symbols. David Tall and Michael Thomas made use of a micro and a cardboard 'maths machine' to encourage young pupils to abstract. They designed a three-week course for a mixed-ability group. There was a pre- and a post-test to look for any effects and these are compared against published research. The results show a positive gain in understanding. The authors feel that this initial introduction provides pupils with some 'mental pictures' they can carry through to later years.

Simulation

Mathematics has long been used as a tool with which to explore and describe the world in which we live. In the past the accepted approach was a deterministic one. The assumption was that there existed some precise mathematical formula lying in wait for the sharp-eyed mathematician. If any errors occurred after data collection and any subsequent calculation then the formula might be discredited or they were explained as human error. Much of what is taught in schools under the umbrella of applied mathematics still reflects this deterministic approach. Probability and statistics usually form a separate course and the stochastic approach to the modelling of observed behaviour has yet to be fully exploited. Mathematical models are after all models and not reality; they express our 'best' explanation to date within the assumptions of the model. There may still remain some elements within it which we feel we are unable to explain other than putting them down to chance. Indeed the whole behaviour might be stochastic.

Now that the micro is available with its inbuilt random number generator we can begin to explore 'random' behaviour through a simulation. Simple situations are readily simulated with short programs; the rolling of dice, the tossing of coins and the drawing of numbers from a hat. These provide quick data samples through which many situations can be explored. Young children can see if a six is really harder to get or not. How long must I wait for two sixes or even two in a row? Do three heads in a row appear as quickly as two heads and a tail when tossing a coin? If I select numbers from 1 to 10 and total them as I go along, how long does it take me to get a score of 40? Does not replacing the numbers make any difference? All these can be conducted without a micro and at the initial stage should be. However, the need for more and more data makes the activity less of an inquiry and more of a chore. When data is generated by the micro the pupil is liberated to think more about interpreting behaviour and less about gathering data. These simple situations can then set the scene for more complicated ones.

The important fact that a program controls the computer simulation can easily be forgotten by children as they become wrapped up in their responses to some dynamic system. These often involve pupils in a decision-making process. These valuable teaching aids can encourage pupils to interpret information and so make an informed judgement. They may offer pupils a number of variables which they are invited to control, providing them with an opportunity to develop their own strategies. Alternatively the challenge may be to find out about the model that the programmer has used in the simulation. This black box approach is after all close to the situation that faces the modeller whose own black box is the real world. Whilst the pupil may lift the lid and list the program there is no such check for the modeller.

Joe Watson points to the recent growth of interest in probability and statistics in our schools' mathematics curricula and considers how simulations on a micro may be used to help pupils gain insight into stochastic behaviour. The situations he considers range from the rolling of dice to road traffic control, radio-active decay and stock control.

Investigations et al

The National Criteria for mathematics and the Cockcroft report both refer to investigational work, rather than investigations. Unfortunately, the latter term is now more commonly used and is often interpreted to mean that there are these self-contained activities that require a different teaching approach, but that do not influence what we do the rest of the time. Problem-solving is another activity which it is thought we must conduct separately, and yet there is often a difficulty in distinguishing the difference between the two activities. Add to this the need for our pupils to undertake extended pieces of work and we have a confusion about just what they should do and when they should do it. If we view this holistically, the emphasis changes from checking that we are indeed covering each of the three components separately to recognizing that through changing our overall teaching approach we are ensuring that all these and other activities can take place in our classroom where and when appropriate. A problem-solver and an investigator might be working side by side, each stimulated by the same initial task, and each sustaining their efforts and producing their own extended pieces of work. There is a real danger that if we compartmentalize these activities and indentify our own carefully selected stages of development within each with the expectation that our pupils will all follow these paths, we will lose the essence of investigational work. This must be in encouraging pupils to ask their own questions and in helping them to formulate their own answers. The 'what happens if' question demands our answer most when we ask the question ourselves.

In his look at the role of the micro in developing mathematical investigation, Derek Ball asks whether software designed around an investigation can often remove the essential investigatory process. By supplying all the necessary stages for the user much of the decision-making is removed. More appropriate software should allow the user the opportunity to pose his own questions, make his own decisions and test his own hypotheses. Derek considers the 'what happens if' question to be the most pertinent to any true investigation and this must be a necessary element to any supporting software.

Ron Taylor takes a close look at the way simple ideas can be presented to children who then develop sufficient confidence to be able to extend these ideas for themselves. He relates his own classroom experiences and offers examples of pupils' work. The micro acted as both initiator of and supporter to the pupils' activities and allowed them to develop their own ideas successfully. Ron takes a close look at the National Criteria for mathematics and demonstrates how these are met from this style of work in the mathematics classroom.

Algorithms

During our school lives we are shown a wide variety of methods which we dutifully try to carry out on set exercises. We met them for long division and multiplication, finding percentages, factorizing quadratic expressions, finding maxima and minima and so on. Very often these would be demonstrated with some well chosen examples and the general procedure would be extracted from them. Only occasionally is the method presented as a sequence of instructions. For example, to find 15 per cent: first find one-tenth of the quantity; then halve this answer; add these two results together. This is the method most people seem to use in practice, despite the usual school textbook method — put 15 over 100 and multiply by the quantity over 1 — that so much teaching time has been invested in. The sequence of instructions that encapsulate the method which provides us with the way of arriving at the solution to our problem, constitutes that particular algorithm.

Rather than just learning how to implement standard algorithms, the micro enables the emphasis to be placed on their construction and exploration. It provides children with the chance to test their own algorithms which they have formulated in the available programming language. Our everyday language, which is rich and full of ambiguities, sometimes lacks the necessary precision required, whereas the formal structure of a programming language provides a more suitable medium of expression. You might like to try and write an algorithm for finding compound interest at variable rates over variable time intervals, using everyday language and then in, say, BASIC. This algorithmic approach enables and encourages children to develop a clear understanding of the mathematics they are using. It is akin to the adage that one way of finding out whether you understand something is to try to explain it to someone. Here the something is mathematical and the someone is the micro.

Adrian Oldknow looks at the algorithmic approach in greater detail. He offers the reader a look at how some familiar situations in mathematics, which may already have standard algorithms, can be introduced to children in unusual and unexpected ways. Adrian emphasizes that teaching with this approach does not mean that what is required is a new set of algorithms for implementation by pupils but rather that pupils should be presented with situations which allow them to construct, try out and develop their own.

David Johnson raises a number of issues relating to anyone involved in a study of mathematics, the role the micro can play in such a study and the environment in which it takes place. He concentrates on some of these and through considering particular examples, highlights his views on 'dynamic procedures'. This is when a concept like 'primeness' can be thought of as procedure for determining whether a number is prime or not. This procedure can be constructed by children as a computer program which requires precision for it to work successfully. It may take a number of attempts to get it working correctly, but the rigour this entails can be thought of as analagous to 'mathematics as a formal system'.

The future

Progress will not cease, indeed the pace at which developments are made is likely to accelerate. Will new technological advances make an encumbrance of much of the mathematics we currently generate through our use of micros, no longer providing us with the justification 'because it's around the micro'? With the advent of more sophisticated machines using higher-level commands and more direct forms of input like the clever 'mouse' that we just push around or through recognition of the spoken word, we may begin to question what there is for us to teach. It is impractical to try to meet such changes by just updating and redesigning syllabuses with different mathematical content. The content must take second place to the underlying processes that are involved when true mathematical investigation or problem-solving takes place. To see a group of pupils huddled around a micro hotly debating the pros and cons of their respective ideas is to see a group of researchers in action. It is important that we create the right environment for our pupils, where they can begin to take the responsibility for their own learning and where the learning of mathematics is promoted for its own sake. That way we are equipping our pupils with those skills they require in order to adapt to the rapid changes that take place within our society. It is indeed a case of trying to encourage pupils to 'learn to think for themselves', despite Orwell's prophecy!

A microcomputer in every mathematics classroom

Richard Phillips, ITMA Collaboration, Shell Centre for Mathematical Education, University of Nottingham

There are many ways of using a computer to teach mathematics; and there are many people who will tell you that there is only one way, their way. But what puts microcomputers apart from other educational equipment is the immense versatility of the device. It would be tragic if we commit ourselves too early to one way of working. This article looks at the teaching possibilities of having a single microcomputer in a classroom. This is an especially interesting arrangement: like many ways of using a computer, it can be exciting, visual, and interactive; it stimulates discussion and sets challenges; it gives rapid feedback and gives opportunities for personal expression. But unlike computer use by individuals or small groups, it allows the full participation of the teacher.

Here is a lesson which involves a computer, a class and a teacher.

A lesson about decimals

We begin with a transcript of the first few minutes of a lesson about decimals which uses a program called ZOOM. A whole class of mixed-ability first-years (11/12 years) are sitting in front of a large monitor. The teacher is the keyboard operator.

'Right, now I want you to have a look at this.' The teacher pauses to get everyone looking at the screen. 'Can you see that line? The top arrow is filled in. The bottom arrow is empty. It means the number that we have to guess is less than ten and more than — what?'

'Nought'

'Nought, all right. So I want a guess. I want to know what you think the number is.' There is quite a long pause as, one by one, hands go up.

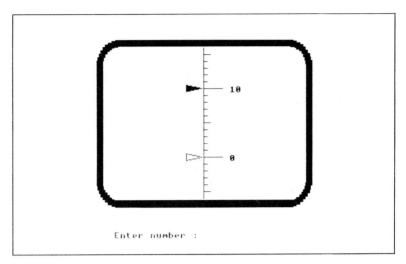

Figure 1. ZOOM before guessing a number between 0 and 10

'Karen?'
  'Five.'
  'OK, five, let's have a try. Let's put five in and see what it says.
Now what do you think that means?' Several pupils speak at once.
'Alex?'

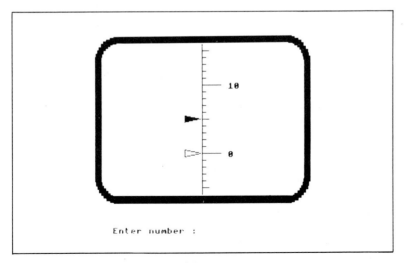

Figure 2. ZOOM with a guess of 5

'It's lower than five . . .'

'. . . but still more than nought. Would anyone like to guess again? Yes? Rebecca?'

'Three.'

'Now what does that tell us, Joanna?'

'More than three and less than five.'

At least half the class have their hands up now, but the teacher picks someone whose hand is not up. 'So what do you think it is, Adrian?'

'Four.'

The teacher enters four and many of the class are looking puzzled. After some reflection, a few hands go up.

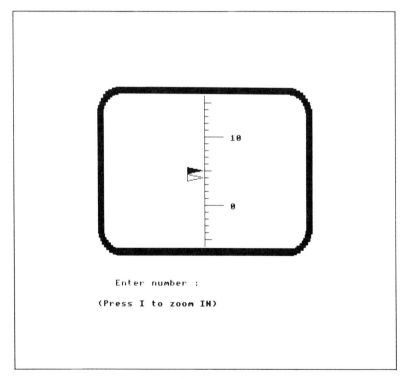

Figure 3. Zoom with a guess of 4

'So where is the number definitely then?'

'It's four.'

'It's definitely four? What does that empty arrow mean?'

'It's higher than four.'

'So what do we know about this number now?'

A brighter pupil comes in with the suggestion, 'It's point something', but the teacher chooses to ignore this.

'What is it between?'

Someone says 'Three and four' but another pupil corrects this to 'Four and five'.

'And there is something on here. It says "Do you want to zoom in?". Let's see what happens.'

The graphics change as if a camera is zooming in on the number line. Everyone watches with great concentration.

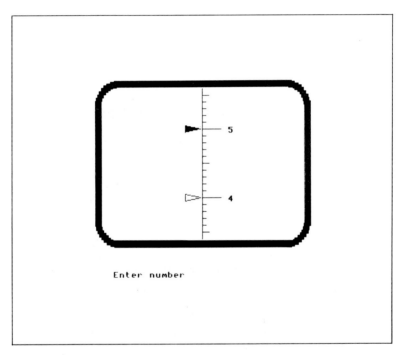

Figure 4. ZOOM with 'zoom in' graphics change

The teacher takes it very slowly, with long pauses between each sentence. 'What are we going to do? . . . Do you know what those numbers are? . . . Put your hands up — don't just shout out please . . . Or make a guess . . . If you want, make a guess . . . Make a guess, Zoë?'

'Decimal point.'

'Decimal point? Well make a guess at a number then.'

Quite a few hands are up now, but the teacher is still prepared to wait longer to carry more with him. 'Anybody? Do we know a number between four and five? Any of those places on the scale? A number? Kate?'

'Four and a half.'

'Four and a half. Do you know which four and a half is? Would you be able to point to it?'

'It's half way between four and five.'

'That one? That one there. Does anyone know another name for that? Because you can't really type in "four and a half". What else?'

'Four point five.'

'Four point five. Let's try that one. Now what do we know? . . .'

The activity continues. After zooming in a second time, the next guess hits the target number: 4.15.

They start again with a new number. The pace is faster now, but the teacher is still prepared to slow down with plenty of time for reflection when he feels it is appropriate. As might be expected, there is a great divergence in pupils' level of understanding, but it is quite clear that it is an activity that everyone is prepared to work at.

The second game ends with them finding a number with three decimal places. Instead of beginning a third game, the teacher gets them to split into groups and continue the activity without the computer. One person in each group is the 'computer' who thinks of a number with three decimal places. Others in the group make guesses and the 'computer' responds 'too high' or 'too low'. It takes very little explanation to set this in motion. Groups work at their own pace. Some choose to write the guesses down, while others are happy to do it in their heads. One of the weaker pupils seems to find it helpful to copy the style of the screen graphics in ZOOM to record the guesses. A bright group increases the number of decimal places with every new number. This work in groups continues up to the end of the lesson.

Although there is no such thing as a typical lesson with a computer, this activity with decimals illustrates a number of common elements of teaching with a computer. Let us look at this lesson from several aspects: its use of computer graphics, how it offers pupils time for reflection, the varying roles of the teacher and pupils, and the way this style of computer use affects classroom management.

Computer graphics

Many educationalists underrate computer graphics. They use the half-derogatory phrase 'electronic blackboard' and talk about pupils 'gawping' mindlessly at the screen. Of course, it is easy to look at a picture or graphic display without learning anything at all, but then most of the words that we read and hear will pass through us without leaving a trace. If they are used well, computer graphics are a powerful aid to learning. It is helpful to distinguish two main ways that graphics help us to think: they help us while we are looking at them by setting out information in a clear, easily assimilable way, and they also help us if they provide us with images which we can continue to carry in our heads after the graphic display has disappeared.

The lesson with ZOOM does both these things. While the activity is in progress the screen sets out the current position in a very clear, simple way. It is like the board in a board game. Without it pupils would be burdened with a number of minor cognitive loads such as the need to remember what numbers have been tried, or the need to make notes. The aim of this opening part of the lesson is to expand pupils' broad understanding of decimals, and to stand any chance of achieving this, any unnecessary mental burdens need to be removed.

The graphical display changes in step with the activity, in the same way that moves are made on a board as a game progresses. This kind of graphics, which changes to follow the argument closely, is called 'progressive graphics'. To some extent, blackboards and overhead projectors can also be used to show progressive graphics, but they cannot match the computer screen in its effortless adjustment to each step in the activity.

In this lesson the computer graphics are not only useful while the activity is in progress. The image of the number line may be retained long after the lesson is over, and should provide a powerful way of thinking about decimals. When the program 'zooms in', the animation lets us imagine ourselves moving closer and closer to the line. There is some evidence that people find it easier to visualize actions than static images; in other words it is easier to visualize yourself doing something, like moving a camera closer and closer, than to visualize a static picture of the number line. The use of animation in ZOOM seems well judged to producing this effect.

Reflection
Someone can only understand a complex topic like decimals if they are given the opportunity to reflect on the ideas involved. Understanding does not all come at once and it is clear that there are many aspects to understanding decimals and that not all children will acquire these in the same way (Swan, 1982). No one really knows the precise way that this kind of understanding is

acquired but it is probably brought about by concentrated attention followed by some personal insight or discovery. It is not a process that anyone can control, but it is possible to create an environment where difficult ideas are likely to be grasped. One of the simplest manipulations is to give pupils adequate time to think: there are a lot of pauses in the lesson described. The teacher asks questions and is prepared to wait for an answer.

Some of the questions create a kind of conflict in pupils' minds. For example, when the class were asked for a number smaller than five but bigger than three, most were confident that the answer could only be four. The revelation that the number is bigger than four creates a cognitive conflict. As rational beings, we feel uncomfortable with this kind of contradiction, and it creates a climate where we will often work hard to resolve it.

For some people, working in a group is an important aid to understanding. Social pressures tend to keep the focus on the important issues. Competition sometimes helps, but the satisfaction of cooperating as a group is probably more productive than competition.

Does a teacher working with a computer help to create this kind of fertile climate? Teachers who regularly work in this way report that it does, and the ITMA Collaboration, who have observed several hundred lessons in this style, have accumulated evidence that the introduction of a computer will cause classes to adopt different ways of working. This style of teaching seems to bring together a number of helpful elements: the teacher's strategies of questioning and silence, the group dynamics, and the computer's power to remove unnecessary cognitive loads and present information in a way that is easy to assimilate.

The computer seems to support the teacher's power to use questions effectively. If the class had to guess a number inside the teacher's head, rather than one inside the computer, it would be hard for him to discuss the problem also in a detached way. Questions like 'So what do we know about this number now?' would be almost impossible. Because the computer is doing most of the work of managing the activity, the teacher has more time to think about the questions he might ask. After asking a question, it is possible to maintain a longer silence when there is a television screen to look at. For a number of different reasons, working with a computer seems to encourage opportunities for reflection.

Classroom rules
In the lesson with ZOOM there is a sense in which the teacher and class are working together against the computer. The teacher does

not know what number the computer has set. He could, quite legitimately, make a guess himself. In this role he would be behaving as a fellow pupil. He might also choose to spend some time in a passive role, perhaps doing no more than being the keyboard operator for the class. Computers seem to have a habit of disrupting traditional classroom roles in a way which adds variety to a lesson. At the end of the lesson about decimals, some pupils were required to take over the computer's role. This was something which they did without difficulty and in a quite natural way.

The main role of the computer in this lesson is as a 'task setter'. As has already been mentioned, the teacher has much more freedom in the kinds of questions he can ask when it is the computer, rather than the teacher setting the task. It also seems a more exciting challenge to guess something concealed inside a computer, than to guess something inside someone else's head. The computer can set its own rules: no one questions that '4.5' is acceptable to a computer while 'Four and a half' is not. But if the class were playing against the teacher rather than the computer, they might well have seen this as unnecessary pedantry.

Issues of classroom management

An interesting question is: when during the lesson do you use the computer? Often it is inappropriate to use it for the whole lesson. Teachers who are new to this way of working will often 'save up' the computer to use it as the end of the lesson. They may feel the class must be adequately prepared before it is brought into use; they may fear a big drop in motivation if the computer is switched off halfway through the lesson, or they may see it as a useful 'carrot' to offer when other work has been completed. Although there can be no firm rule about these things, these arguments are usually wrong. Some of the most effective lessons we have observed have begun with the computer and then led into other work. The explaining power of the computer is considerable. At the beginning of the lesson with decimals, the teacher used just 44 words of explanation, and it appeared that everyone understood the rules of the activity. This explaining power was also evident when the computer was switched off and work continued in groups. The teacher had to say very little: it was clearly understood how one in each group copied the computer and the others were players. Motivation does not seem to drop once the computer is switched off: the computer will often generate a high level of interest and involvement which continues through to the end of the lesson.

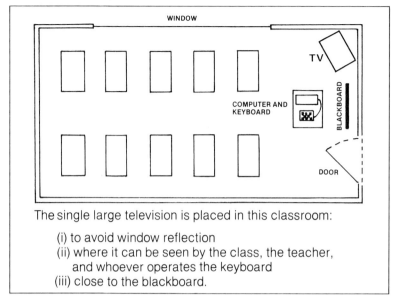

The single large television is placed in this classroom:

(i) to avoid window reflection
(ii) where it can be seen by the class, the teacher, and whoever operates the keyboard
(iii) close to the blackboard.

Figure 5. Microcomputer positioned for whole-class work

The main problem with using a computer as a whole-class teaching aid is the difficulty in getting everyone to see the screen clearly. The problem is soluble without a great deal of expense if some care is taken. It is a matter of using the largest television or television monitor you can, of taking the trouble to adjust it properly, and of positioning it well in the classroom. When there is a choice, use software which is designed well for this purpose, and identify pupils who may need to move closer. It is unfortunate that local education authorities are not more sensitive to this problem: there is a lot they could do to help, if only in giving the right advice.

Some other lessons

This discussion is not based on just one lesson with decimals but on the observation of many hundreds of lessons by members of the ITMA team. We have seen classes work with many different programs and we have watched teachers with a wide range of teaching styles successfully adapt to working with a micro. Here are two more examples of this style of working.

SALESMAN
The idea behind this lesson is to compare human problem-solving with problem-solving with a computer. The class tackle a simple

version of the travelling salesman problem. They agree on where the salesman lives and on ten places he must visit. With the aid of maps and mileage charts, each group tries to find the shortest route which will take the salesman from his hometown to each of the places in turn, and back again to his hometown. The activity is mildly competitive: a record is kept on the blackboard of the best route found so far. A short route found by one group may be turned to advantage by another group who modify it and make it shorter still. Discussion may focus on strategies. Should you try to visit the most distant place first? Should your route 'go round in a circle'?

```
Aberdeen         Glasgow          Northampton
Aberystwyth      Gloucester       Norwich
Barnstaple       Guildford        Nottingham
Birmingham       Hereford         Oxford
Brighton         Holyhead         Penzance
Bristol          Hull             Perth
Cambridge        Inverness        Plymouth
Cardiff          Kendal           Preston
Carlisle         Leeds            Salisbury
Carmarthen       Lincoln          Sheffield
Colchester       Liverpool        Shrewsbury
Dorchester       London           Southampton
Dover            Maidstone        Stoke
Edinburgh        Manchester       Stranraer
Exeter           Middlesboro'     Taunton
Fortwilliam      Newcastle        York


Where does the salesman live?
```

Figure 6. Screen display in SALESMAN

The computer is available at this stage as a calculator to check the length of routes. It is also helpful to have hand calculators available.

After some time everyone comes together. In discussing what has happened the teacher asks, 'Have you found the *shortest* route?' Clearly there could be a shorter one. The class then look at the computer attempting the same problem.

The computer begins with an exhaustive search. The computer works through every possible ordering of the ten towns; each remains on the screen for a moment as its distance is checked. Even though it works quickly, it has 3,628,800 (or 10!) routes to check and it will not be finished until long after the end of the lesson. Is it a good way of finding a route? Can we be sure it will find the shortest route? How is it different from the way we did it?

32

This is interrupted and the computer tackles the problem in a different way. The 'three-cut method' shortens routes using a kind of sorting procedure. This produces a level of performance which is very similar to the class's own. It is much faster than an exhaustive search but it doesn't guarantee to find the shortest route. After trying this for a few minutes, it may find the shortest route discovered by the class. It may even discover a shorter one.

What are the differences between the way we solve this kind of problem and the way a computer solves it? Here are some comments made by pupils in discussing this at the end of the lesson.

'We can do it quicker because we see it on the map. The computer tries every possible route.'

'When we do it, we can use our common sense to see which looks the shortest route. The computer just shuffles them in all different directions.'

SUNFLOWER

One of the most natural ways of using a computer is to strive to discover some kind of knowledge which is concealed inside the machine. A mystery concealed in a box is archetypical. It goes back to Portia and Pandora, and in recent times it surfaces in countless ways: in adventure games, in black boxes, in 'phone phreaking', and in the way that 'computer nuts' tease out the details of how their computer works. The lesson with ZOOM showed how a secret concealed inside a computer is very different from a secret inside a teacher's head.

One successful variant of this idea has been the science simulation program where the computer simulates a scientific system and challenges you to discover enough about the system to control it or predict it. For example, the computer may simulate the ecological balance in a pond, or the manufacture of sulphuric acid in an industrial plant. Here is a mathematics lesson which uses a kind of science simulation program. The interest is not so much in the science as in the method of discovery. It is a lesson about systematic investigation, which also brings in decimals.

A class of average-ability 12-year-olds read the instructions from a large television screen:

'According to the Guinness Book of Records, in 1983 Martien Heijms of The Netherlands grew a sunflower with a height of 7.38 metres. This is believed to be the tallest ever grown. Can you beat this record? You have three chemicals A, B and C which are added to the plant's water in any amounts you wish. Can you find the concentrations of A, B and C that will grow the world's tallest sunflower?'

One pupil reads '7.38' as 'seven point thirty-eight' and this is corrected by another. Someone says they need 'compost and water' but the teacher points out they don't know what the chemicals are. The teacher asks, 'How shall we go about this?' but gets no suggestions. The class agree to make a start by trying 5mg/l of chemical A, 10mg/l of B, and 15mg/l of C. The sunflower wilts and dies. Someone says the amounts 'might be too much', another is concerned that 'one's higher, one's lower' and suggests 10mg/l for all three chemicals. Again the sunflower wilts and dies. The teacher continues to enter suggestions from the class with very little comment. Over a ten-minute period the whole class argue about what should be entered. They try six different combinations of chemicals, every one of which causes the sunflower to wilt and die.

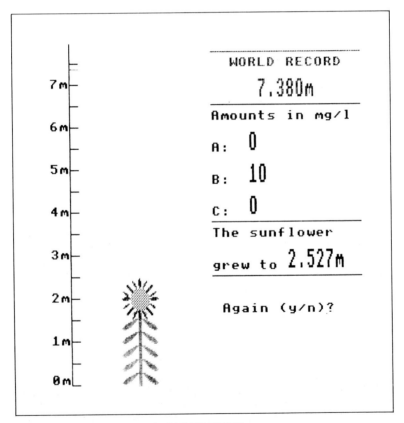

Figure 7. Screen display in SUNFLOWER

Despite this apparent failure the whole class is involved and everyone seems to have an opinion about what should be entered next. It would have been easy for the teacher to intervene during this period and drop in a hint to get them going, but he is prepared to wait. Eventually the breakthrough comes. A pupil says 'Put nought in of the whole lot', and although this idea is dismissed by several other pupils, the teacher enters it, and rather than dying, the sunflower grows to a modest 1.247m. There are cheers!

The teacher briefly takes over. Everyone is asked to write down the result and he asks 'What does that tell us?' A few pupils draw the conclusion that only small amounts are needed. To test this they enter 0.5mg/l for each chemical but once again the plant dies. The second breakthrough occurs a minute or two later when another pupil sees that the effects of the chemicals must be investigated separately. She suggests 0.1mg/l for A with no B or C. This time the sunflower not only grows but also flowers.

This is halfway through a 50-minute lesson. The whole class work as a single team for the entire lesson. The teacher's main interventions are to insist everyone keeps a record of the more important results. The discussion is noisy and almost everyone is involved — halfway through just two pupils show signs of losing interest. Gradually they begin to understand the effect the chemicals are having, and three minutes before the end of the lesson they produce a combination of chemicals which breaks the world record.

Microcomputers as teaching aids

The use of a microcomputer as a teaching aid can only flourish within a school that encourages it. The location of machines within the school is perhaps the most important factor here. If all the computers are kept in a computer room, if they are difficult to move, or difficult to book, no one will use them regularly in this way (see Phillips *et al*, 1984). The ideal is to have one microcomputer in every mathematics classroom: it may become as permanent a piece of equipment as the blackboard. An alternative is to have just a small number of rooms equipped in this way, chosen to make it easy for teachers to swap rooms. Moving pupils is generally easier than moving equipment, although circumstances vary greatly from school to school.

Some uses of computers as teaching aids are of very short duration: an activity lasting just a few minutes may make a very effective contribution to a lesson. But few teachers would go to the trouble of booking or moving a machine for such a short use. Some may even feel some unjustified guilt at letting a machine stand idle for most of a lesson.

It is a question of attitude: if you believe that microcomputers are fundamentally labour-saving devices which should assist busy teachers (or even replace them), or if you believe that someone is only really interacting with a computer when sitting at the keyboard typing, then none of this is for you. But if you are prepared to recognize that there are many good ways of using computers in schools, you may appreciate that the peculiarly three-sided combination of the teacher, the pupils and the computer has some rather special benefits.

References

PHILLIPS, RJ, BURKHARDT, H, COUPLAND, J, FRASER, R, and RIDGWAY, J, 'The future of the microcomputer as a classroom teaching aid: an empirical approach to crystal gazing', *Computer Education* 8, pp173-177, 1984
SWAN, M, *The Meaning and Use of Decimals,* Shell Centre for Mathematical Education, University of Nottingham, 1982

Programs
SALESMAN is included in *Teaching with a Micro: Maths 1.* ZOOM and SUNFLOWER are included in *Teaching with a Micro: Maths 2.* Both are published by the Shell Centre for Mathematical Education, University of Nottingham.

Introducing a micro into a primary school

James Whitbread, Botley Primary School, Southampton

The microcomputer is a constant source of amazement to me. What a neat little gadget it is! In-service courses that I attend always fire my enthusiasm and add to my conviction that all this expensive hardware should be in use every minute of the school day. Our children deserve the sort of organization and approach which makes optimum use of our resources. But what are the realities of the classroom?

My school is a small primary school. We have approximately 180 children, six teaching members of staff and one microcomputer which we have had now for five terms (as in many schools, a very willing PTA is at present raising funds for further machines). We have traditional individual classrooms in two buildings separated by a playground/car park area with lots of kerbs and bumps. Although we have a good quality work station on sturdy castors it is necessary to carry all equipment from building to building. Perhaps because of this the computer is used on a rota basis with each teacher in possession for one week at a time. If for any reason he or she cannot make use of it (because of hall periods, games, etc) it is reallocated to another teacher in that building. There is usually someone eager to use it. We can look forward to having the micro in class once every six weeks, and with half-term or holidays in between, this becomes every couple of months.

Staff expertise is limited effectively to the two members of staff who attended the obligatory two-day introductory course. All teachers have taken the computer home for at least a weekend and can load and run programs. We have no regular in-service training at school but there is interest and we share our experiences.

With this background in mind it is no surprise that the 'intruder' was viewed in two ways according to circumstances — either as a little treat or as an irritation sent to upset the smooth running of the week. Whichever opinion was held the microcomputer for a long time looked like making no positive impact upon the mathematics curriculum.

For the first year of using the computer much of our time was spent dabbling with software which we acquired without there being any real direction to our work. Although we are still finding our feet we all believe that in our present circumstances the micro is best used 'little and often'. When the computer eventually appears in my classroom it is impractical to embark upon some of the lengthy investigations, problems and adventures that are offered to us in many software packages. 'When is it our go?' — 'It's your group next.' (You know very well that that could be the day after tomorrow or it could be in six weeks' time. It seems exaggerated but it really does happen.)

In order to waste fewer of those precious moments I like to restrict my children to tasks which take ten minutes or so to complete and keep my fingers crossed that every child will have several hands-on sessions in one week. The way that we use the computer therefore placed restrictions on what software we chose.

Drill-and-practice programs comprised the bulk of the software available to us right at the beginning. They were used frequently because they could fill up almost any time slot. There was a mistaken belief that a computer being used all the time was being used well, but the programs did give wary staff and pupils confidence and helped to improve everybody's keyboard skills.

The choice of software even at that time was quite bewildering and, as a result of our dabbling, the children must have felt rather uncertain. A need clearly arose for us to stop, think, find a 'core' item of software and concentrate our efforts upon it.

Following a series of evening sessions in which turtle graphics was introduced by an 'enthusiast' I decided that we should look more closely at the version that was readily available to us, DART. It seemed to solve many of our problems. Immediately we could introduce 'computer mathematics' appropriate to our children and classroom organization, and the program was flexible enough to be included throughout our mathematics scheme. DART does require some hands-on preparation, but we set about finding ways of introducing the tool to children across the entire primary age-range.

Teachers of the younger children threw up some initial doubts . . . the numbers are too large . . . it looks too complicated . . . not very interesting for our little ones. These were fair comments at the time but with a little imagination 'fun' activities were thought up for even the youngest infants.

We made wide use of teacher-created procedures to draw on the screen and to control the movement of the pointer and also used stickers (Figure 8) which we put on coverlon and then apply straight on the VDU screen. (Card shops and toy shops sell a wide range of very colourful stickers from dinosaurs to fairy-tale princesses and they are not a great drain on petty cash.) Our children are really motivated by the visual appeal of these.

The program was first tried with infants in its raw form. Some met with limited success but the numbers required to register any significant forward or backward movement (up to numbers in the

Figure 8. Stickers for the screen

hundreds) or turn (up to 360) were clearly too large for the majority. Notation was a major problem with children declaring 'we've got to go 150' and then entering FORWARD 10050!

Four simple procedures were created for the youngest children which it was hoped would enable them to do meaningful work without the need for numbers. The procedures were:

F   (FORWARD 20)
B   (BACKWARD 20)
L   (LEFT 90)
R   (RIGHT 90)

With these the pointer can be moved around the screen in easy-to-see jumps and control is almost immediately in the hands of the child. Now a typical succession of tasks might be:

1. Take the pointer for a crazy walk around the screen.
2. Can you get the pointer as close as you can to the edge of the screen?
3. Take the pointer to the other edges of the screen.
4. Take a walk with the pointer right around the screen.
5. Get the pointer to the teddybear (sticker). (To start with the sticker can be placed directly above, below, to the left of, or to the right of the pointer in its centre position.)

With control over the movement of the pointer it becomes possible to use the program as a problem-solving medium, to reinforce linear measurement, work on shape, number, etc. At a later stage the procedures F, B, L, R can be used with values. If procedure F is BUILT WITH DISTANCE, ie,

BUILD F WITH DISTANCE
MAKE DISTANCE DISTANCE * 10
FORWARD DISTANCE
(press ESCAPE)

then F1 is quite a large step. F10 will take the pointer to the edge of the screen and 20 will be the largest number needed. We can of course scale F anyway we wish. Similarly we can

BUILD L WITH TURN
MAKE TURN TURN * 45
LEFT TURN
(press ESCAPE)

This creates units of turn of 45 degrees.

By adopting this system of scaling up turns and distances it is possible to ensure that children only need to use numbers that they are familiar with.

Our younger juniors were introduced to DART by use of the four elementary commands, FORWARD, BACKWARD, LEFT, and RIGHT (which do of course have to be followed by a number). Although most did not know what degrees were, they very soon found (directly or by the jungle telephone) that LEFT 90 gave a nice square corner. After an initial play period with these commands we set up a series of tasks called 'shepherd'. The series begins with a procedure which draws a sheep on the screen and the problem is to draw a closed fence right around it (so that it can't run away) using the pointer. This could be equally well done using stickers. The problem can be varied by asking for the smallest field that the sheep could live in or the largest. Next more sheep are added by use of the same procedure in different positions on the screen and they must all be fenced in (Figure 9). Eventually by use of another procedure one wolf (or more) is added to the screen and the problem is to enclose the sheep without having a wolf in the field (Figure 10). The children are required to work with more accuracy as they progress.

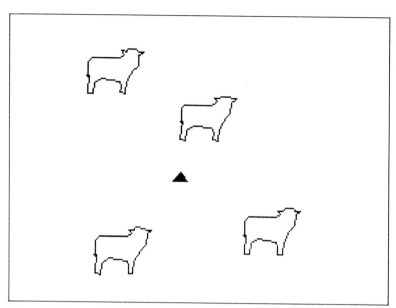

Figure 9. Sheep drawn on the screen

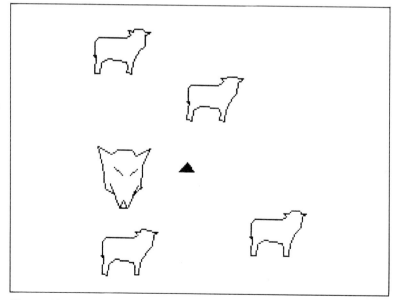

Figure 10. Sheep and wolf drawn on the screen

A second activity for the same aged children uses a procedure named 'track' which draws a simple race track on the screen with the pointer centring on the start line (Figure 11). Children take their driving test by trying to drive the pointer around the track without crashing into the barrier. Accuracy in estimating screen distances and angles of turn obviously show better driving. Hazards introduced to the track liven up the proceedings. L plates and mock driving licences sharpen interest further.

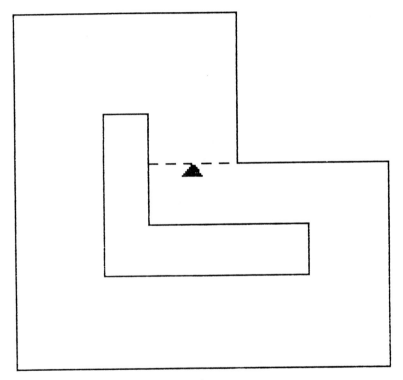

Figure 11. Race track drawn on the screen

Something a little more demanding was needed for the third- and fourth-year juniors and the work that we eventually decided upon was called 'Star Trek' and simulated part of the training of a starship commander. A procedure named 'space' draws on the screen a series of different sized circles representing a 'planet system'. Home planet is in the middle with the pointer neatly centred on its surface. With the screen set up the first task for a trainee starship commander is to make an exploration of the system, visiting all the planets and then making a safe return to earth (Figure 12). Children were eager to try the same task again and again as they discovered different routes and made errors. All commands had to be recorded on a starship log-sheet for analysis by other trainees on return to earth. The dialogue which this sets up forms an important part of the work. The screen display is very close to the sort which the children might see on a science fiction film and it becomes very 'real'. My own class became totally involved with playing this game and its extensions. To make the job of the crew easier we now try to complete the journey by issuing the fewest possible commands. The next problem is to complete a similar journey but to take avoiding action when faced with black holes, aliens, meteorites, etc (stickers or procedures).

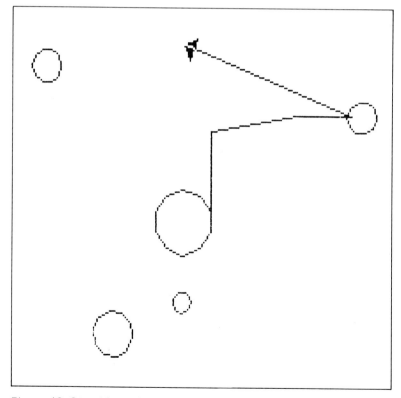

Figure 12. Starship and planets drawn on the screen

At all stages children are urged to make 'soft' landings, ie, to estimate accurately in order to land on the surface of a planet rather than boring into its surface (which might get them demoted). We next focus our attention on fuel economy. Fuel is represented by units of forward and backward movement. With a given amount of fuel children must investigate how many other planets they can visit, being careful to leave enough fuel to get back home; eg, what journeys could you make using 400 units of fuel? They must also compete to find the least possible fuel needed to complete a safe journey to all the other planets and back to earth. The children can always find ways of improving their own performances and nag to get back to the machine. In the latter stages of 'Star Trek' our children are shown how to save their journeys on disc as procedures. This enables them to display one route on the screen and use it for comparison while working on another. It enables them to compare and discuss with other groups or individuals.

Gradually we are becoming more adventurous and our use of DART has progressed throughout the school. Pupils are enthusiastic about it and it has helped create situations in which they can begin to pose their own questions and develop strategies with which to solve them. They have become more skilled as independent learners and this has had a knock-on effect into other areas of the curriculum. We are also beginning to look closely at our existing mathematics programme to see how we make further use of our micro. One such development took place when I was considering how to teach tessellations to my third-year juniors.

The HMI discussion document, *Mathematics 5 – 11*, supports the inclusion of tessellations by saying that 'Through practical experience of . . . plane-filling tessellations . . . Children should come to recognize the geometrical properties of common two-dimensional shapes'. In its detailed catalogue of experiences for children it includes: 'Making patterns from objects or shapes'; 'Exploring two-dimensional space: covering a plane surface with different shapes — with overlap and without overlap'; 'Patterns based on simple shapes, without overlap. Looking at mosaic patterns'; 'Classification of two-dimensional shapes'; 'Patterns: tessellations (tile patterns). Consideration of various tessellations . . . of given shapes including . . . more regular shapes. Tessellations in architecture, mosaics . . .'; 'Consideration of the sum of the angles of triangles from tessellation for equilateral triangles. Similar work with tessellations of parallelograms, of hexagons and of regular octagons with squares'.

I had been given a short procedure for drawing regular polygons. I felt that I could adapt this and construct a pattern-drawing program.

Clearly to be of any real use the children would need to control:

(i) the number of sides of the shape
(ii) the size of the shape
(iii) horizontal and vertical spacing.

They would also need to rotate the individual shapes in each row and 'slip' the rows across one another in order to achieve the more sophisticated tessellations.

The first attempt at a program required seven variables to be input in order to produce the pattern:

(i) number of sides
(ii) size (measured from centre to corner)
(iii) horizontal spacing between centres
(iv) vertical spacing between centres
(v) rotation required in the first row
(vi) rotation required in the second row
(vii) horizontal shift between rows

Since I felt that controlling seven variables would be an unrealistic demand on most of the third-years, 'trimmed down' versions were created. In these the shape was set, the size was set, and the turns were set. Three variables were to be controlled: horizontal spacing, vertical spacing, and horizontal shift.

I decided to test the program out with a group of girls from my third year, and recorded their conversation to gauge their response to it. To my suprise they spent one hour on the program; the tape only survived for 20 minutes. They decided to use a square initially and began moving it around the screen. They then set themselves the task of placing one in each corner of the screen. The following represents some of their dialogue at the early stages.

'. . . still too high . . . it was 100 last time . . . put it higher? no, lower, 100 . . . no, we tried 100 . . . I know, try 80, try 65 . . . return . . . yes . . . it's still wrong, its changed . . . try 10 . . . you are not changing the right one, try 80, that's a guess . . . oh, how did we get that, what did we have . . . I don't know . . . ooh ha ha . . . that's what I think we had before . . . yes . . . no, we want it more perfect than that . . . no we don't . . . yes we do . . . let's try . . . no let's try, let's take, just put 30 instead of . . . just put 1,1 . . . it was 1,0 . . . I remember now I changed it, didn't I? . . . oh no . . . great . . . I think I've got it . . . right . . . no, I don't think that's going to be right.'

Later they began to be less haphazard and built upon their previous efforts. The need to record became apparent as this later dialogue shows.

'. . .that's miles out . . . no, we're completely wrong, I'm not sure about this, get it up to what it was, let's have seven sixty-seven . . . I think we're going to have to change A . . . yep . . . easy . . . oh no . . . ten ninety, ninety . . . ten ninety, ten ninety and ninety . . . oh leave it like that, just put ten ninety, just remember it, got a pen? I'll write it on my knee. Let's try ten seventy, the top one's wrong now . . . I reckon we're going to move that one a bit . . . yes . . . right, what did you try? . . . Oh great, brilliant, we want the bottom ones as they were, but put in ten ninety again, it's not perfect but if we move it back a bit . . . no, I'd leave it or else you're going to mess it up . . .'

Paper replaced the knee towards the end and they finally resolved their problem. Other shapes were tried too and a variety of patterns was constructed over the last 15 minutes. The whole activity generated a tremendous amount of communication between the girls. During this time their communication improved, going from isolated words, figures and phrases to more sophisticated statements relating cause and effect. Their estimation skills and strategies were purposefully used. They had set themselves a goal which was within their reach but the length of time it took them to attain it required tenacity and stamina.

Since this initial trial I have had many successes with the program and adapted it to perform a variety of tasks. It has provided a testbed for my pupils and together we have learned a lot about tessellations. Some of their patterns appear below but their thoughts and experiences remain with them.

As a footnote I must add that over the time that I have been using these programs and observing the children working with them I have often been surprised by children who perform much better or much worse than expected. Those labelled as 'bright' sometimes find it harder than the 'no-hopers'. I'm sure that I am as guilty as any in using the traditionally accepted criteria (an ability to learn tables, add, subtract, etc) in assessing the mathematical ability of my children. However, the introduction of the micro into my classroom has encouraged me to reflect on the appropriateness of those criteria.

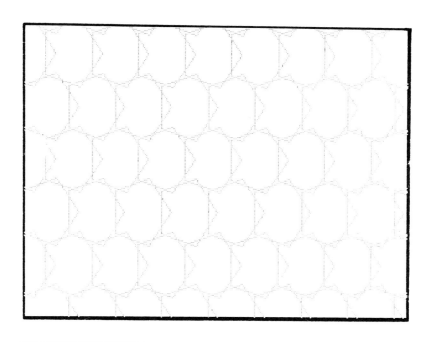

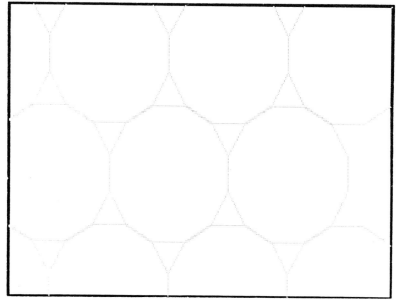

Figure 13a. Tessellations drawn on the screen

47

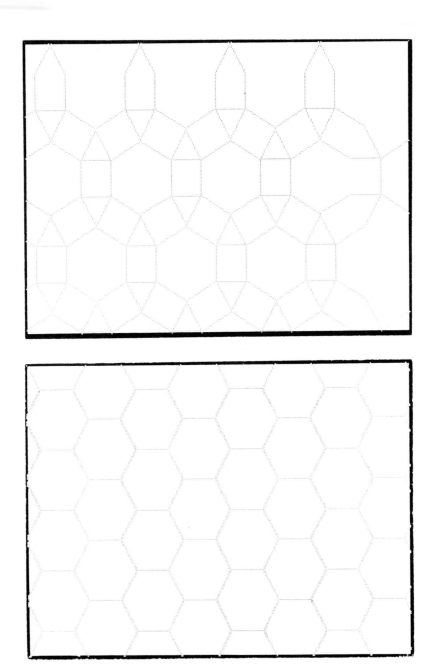

Figure 13b. Tessellations drawn on the screen

48

A micro in my classroom: the elements of algebra

John Higgo, Haileybury College

[The author has taken a major part of this article from one he had published under the same title in *Mathematics in School* 14, 5 (November 1985), and it is reproduced by kind permission. The author has also drawn upon some material circulating within the Mathematical Association's Microcomputers in the Mathematics Curriculum Committee, which he chairs.]

T J Fletcher in his recent DES Discussion Paper (1) points to a new-found enthusiasm among children for aspects of mathematics which are related to the use of a microcomputer. He pleads that 'this enthusiasm should be turned to good effect without delay' and goes on to suggest: 'Secondary mathematics teachers can make extensive use of microcomputers just as they come out of the box. This is because the BASIC programming language with which today's microcomputers are provided is rich enough to cover extensive areas of mathematics, and is easier to learn than many things which are universally established as a part of the secondary curriculum'. It is now a widely held view (2,3) that the mathematical aspects of programming ought to form part of the mathematics curriculum. (In the future the language through which our pupils gain direct access to computer power will not necessarily be BASIC, though.)

Irrespective of future curricula, I believe that algebra (as well as other aspects of school mathematics such as calculus and statistics; cf 4) can be taught more effectively now with the aid of programming ideas. If you are fortunate enough to have fairly ready access to at least one micro during mathematics lessons, it is well worth teaching your second-, third- or fourth-years the elements of BASIC (though you may find that a majority of them already possess adequate familiarity). Certainly, a modicum of time spent on programming should be repaid handsomely in the form of readiness of understanding and motivation.

The main point is that children find a program such as the first one below simpler to understand than a raw formula because it seems to be less abstract, more directly related to their experience — even if the very same formula is embedded in the program!

I find myself increasingly referring to 'how we could instruct the computer to do it' in my teaching of algebra at this level. It is natural for me to turn to this new tool in my armoury, as illustrated by the 'case-study' below.

49

Towards the power of algebra

We are sometimes in danger of compartmentalizing a piece of mathematics. This is detrimental to the learning process, which ought to be child-centred. On the whole, the distinctions between, say, functions, formulae, graphs, equations, numerical methods, programming and problem-solving are artificial and helpful only to the teacher. What follows seemed to a group of 13-year olds a natural progression to the use of algebra, meeting and using some powerful, but comprehensible, mathematics along the way. A computer was used only where it seemed relevant — strategically it provides a path to constructing and understanding formulae. There are many possible variations on this theme and it may be applicable at many different levels.

Lesson 1. I want to hire a Mini for the day. Rentamini charges a flat 25p per mile. Hiremee want £5 for the day plus 10p per mile. Whom shall I favour with my custom?
We soon have some calculated costs, as the table shows.

No. of miles	Cost in pence	
	Rentamini	Hiremee
M	R	H
0	0	500
100	2500	1500
50	1250	1000
. . .	. . .	. . .
. . .	. . .	. . .
33	825	830
34	850	840
33.5	837.5	835
33.4	835	834
. . .	. . .	. . .

What mileages make Hiremee cheaper? What shall we try next?
If the teacher is lucky (or sly), 'why don't we draw a graph of the costs?' So we use the calculated values to plot first R against M and then, on the same axes, H against M. By the end of the lesson they have all found that the critical mileage is somewhere around 33 or 34 (from the graph and one or two have actually started to refine their answers by extending the table, using calculators, as above.

50

Lesson 2. This is getting a bit tedious. Why not get the micro to do the arithmetic for us? Show them how to write a program to find R.

```
10 INPUT M
20 LET R = M * 25
40 PRINT R
```

Explain that this means

Take M →— X 25 >→ R

To find H there are two steps and their order is important.

M →— X 10 >→ Y →— + 500 >→ H

(What keys do you press on your calculator?)

Some do write their programs in two steps.

```
30 LET Y = M * 10
35 LET H = Y + 500
```

Ten minutes before the bell I call a plenary session. What we really want printed out is M, R, H, together as in our tables; so we weld the programs together on the blackboard, ending up with:

```
10 INPUT M
20 LET R = M * 25
30 LET H = M * 10 + 500
40 PRINT M, R, H
50 IF R = H THEN PRINT "EUREKA!"
60 GOTO 10
```

Lesson 3. The pupils work at different computers in pairs, using this program. To my surprise it takes the whole lesson and nine decimal places before any one suggests that the answer is 33.333 . . . recurring! The inability of some to interpolate a decimal between two given decimals is quite revealing. The class has learnt something about systematic experimentation.

Lesson 4. How do you write the two formulae on lines 20, 30 in algebraic notation? So, when R = H we have 25M = 10M + 500. Within ten minutes we are all convinced of the power of algebra.

A few thoughts
1. It is best to introduce an idea like this via a simple practical problem (preferably not distance, speed and time). They would find the whole process more difficult if we started with
'Solve 25x = 10x + 500'.

2. Should first attempts at solving such equations not be numerical in any case? Before teaching pupils to collect like terms by the handful should we not try to answer that cry from the heart of many adults: 'Yes, but what is x?'

3. Any equation of the form $f(x) = g(x)$ can be solved using a numerical method like the above. In the case of $x = g(x)$ we are right into one of the examiner's favourite iterative methods in the sixth form.

Elimination by substitution
Although the synthesis of lines 30 and 35 above into a single formula seemed to come impressively naturally to many of the pupils, either I then forgot about it or the full significance simply did not strike me at the time, even when they carried out both the same kind of process and the reverse (analysing a composite function into its constituents in order to find the inverse) in subsequent lessons. Two terms later they came up against the question.

'Make I the subject of the formula $V = IR$. The power consumed, W watts, is given by $W = IV$. Combine the two formulae to find W, first in terms of I and R, then in terms of V and R'.

All I needed to do by way of explanation was to put in line numbers

```
10 I = V/R
20 W = I * V
```

and the pupils realized that they already knew how to carry out elimination by substitution.

The availability of cheap microcomputers raises far-reaching curricular questions, worth discussing at some length, as to what method(s) of solving simultaneous linear equations we shall want to teach to whom. Without going into the issue at all I shall simply assume, quite reasonably I hope, that there will always be a clientèle below the age of 16 for learning the substitution method. The Hiremee/Rentamini example above points the way to a suitable introductory approach to this topic.

Suppose, for example, that the following pair of equations has 'arisen' from some suitable practical context (such as those found in typical school textbook exercises on linear programming):

$$x + y = 10$$
$$5x + 2y = 30$$

Attempts could be made to home in on the solution in the same way as with Hiremee/Rentamini;

```
10 INPUT X
20 Y = 10 − X
30 IF 5 * X + 2 * Y = 30 THEN STOP
40 GOTO 10
```

The necessary change of subject on line 20 should not be difficult to motivate. Now, shortening the program by combining lines 20 and 30 is not only tempting but essential if the solution is to be found algebraically as in the car-hire experience.

Examples amenable to this kind of treatment abound. A non-linear example with which I have had some success is the open box problem, given as an illustrative example of how treatment of a topic can be adapted to suit different ages and abilities in the latest HMI 5 – 16 document (5).

Percentages and interest

A term earlier I had been trying to instil some of the concepts underlying algebra, eg, inverse functions, into the same class of 13-year-olds via, among other strategies, the micro. I happened to set them a multiple-choice revision exercise from SMP Book G, including the following question.

A man's annual salary of £1000 is increased by 10 per cent and then a month later the new salary is increased by a further 10 per cent. His annual salary is now (a) £1100 (b) £1200 (c) £1210 (d) £1250?

Needless to say, they all chose (b). I spent several of the ensuing lessons sorting this out and enlarging on it. First, I reminded them that

and that we were required to repeat this process, as indicated by the looping arrow. This gave us the program

```
10 INPUT A,P
20 LET I = A * P / 100
30 LET A = A + I
40 PRINT A
```

They readily agreed that we needed to add lines

```
5  LET C = 0
15 LET C = C + 1
35 IF C < 2 THEN GOTO 15
```

53

A dry run helped drive the point home that the increase, I, is not the same both times. (As far as I was concerned, the whole point of the exercise was to do the dry run, but we did 'check your answer by computer' in the next lesson.)

We next considered the question of whether

| decrease by 10% > | is the inverse of | increase by 10% > |

We already had inverse function pairs programs such as

```
10 INPUT X, K          and      110 INPUT X, K
20 LET Y = X + K                120 LET Y = X − K
30 PRINT Y                      130 PRINT Y
```

Why not do the same for percentage increase and percentage decrease? We already had our program for percentage increase. The pupils soon spotted that percentage decrease required only one change in line 30. We did dry runs, of course, and I set the subsequent problems in suitably commercial contexts. And it became clear why decreasing by 10 per cent does not 'undo' increasing by 10 per cent. We were also in a position to answer such questions as:

Which is better for the purchaser if he is offered a discount of 30 per cent: first add on VAT at 15 per cent and then take off 30 per cent, or take the discount off first? (Many adults are surprised at the answer!)

Next we worked, mostly via small amendments to the programs and doing dry runs with the aid of a calculator, on questions such as the following.

1. My wife has just made me insure my life for £20,000. If I die in ten years' time what will this be worth to her in real terms, assuming inflation continues at 6 per cent? Change a few variables . . .

2. A car depreciates at 10 per cent per annum. It costs £6000. What will it be worth in 1990?

3. Find the value of a £10,000 investment at 10 per cent compound interest after 5 years. And at 10 per cent simple interest?

In the middle of this I decided it was about time they all understood FOR . . . NEXT loops, so we modified our compound interest program to:

```
10 INPUT A, P, N
15 FOR YEAR = 1 TO N
20 LET I = A * P / 100
30 LET A = A + I
40 PRINT YEAR, I, A
50 NEXT YEAR
```

I felt it important to keep lines 20 and 30 as two separate steps for these pupils at this stage. It is useful to print out I as well so that they can check their dry runs more readily; they can understand the meaning of 'interest'; they can see how it increases each year; and, especially, that it is constant in the case of simple interest. (To change to simple interest all you do is to change the number of line 15 into 25 so that the interest is calculated outside the loop.)

Increasing but limited

The investigation outlined below actually went down very well with my fourth-year O-level set. This further illustrates some of the points made earlier, stresses the fact that the numerical approaches which I have been advocating sow the seeds of ideas about convergence and sheds further light on the power of a computer as a learning aid.

Banco de Buenos Aires et Tel Aviv offer you 100 per cent pa interest. Suppose you invest £1. How much will it be worth after one year? Ah, but suppose they work out the interest owed every six months and then give you interest on the total amount for the second six months? Show that, if the interest is compounded quarterly (ie, worked out and added on every three months), your investment would amount to

£(1 + 1/4 + 5/16 + 9/32 + 45/128)

after a year. [They have long since reached for their calculators.] Guess what you would receive at the end of the year if the interest is compounded monthly. Do you think we could bankrupt the bank by asking for the interest to be compounded weekly?, daily?, every second? . . .

Write an algorithm (or flowchart or spreadsheet or program) to find the final amount when the interest is compounded monthly. RUN it.

[Or perhaps develop the program as a class effort:

```
10 LET A = 1: T = O
20    REPEAT
30       LET A = A * (1 + 1/12)
40       LET T = T + 1
50    UNTIL T = 12
60 PRINT A                    ]
```

By adjusting the program, plot a graph of A against n, the number of times per year that the interest is compounded. What is really happening? At what stage is computer error coming into play?

Use the e^x button on a calculator to find a value for the irrational number e.

Experiment with changing (i) the initial sum (ii) the rate of interest, and try to predict the effect. Test your hypothesis. Generalize . . .

Points arising
1. This example shows 'how at a lower level of abstraction the computer can give an insight into more formal concepts. In this way ideas may be presented practically to less able pupils and prepare more able students with appropriate intuitions for later formalizations'. (3)

2. It is possible that much of what is now taught in the sixth form may become suitable material for a wider range of ages and abilities.

3. Simple algorithms together with insight into the problem should be preferred to mastery of half-understood sophisticated methods. Simple algorithms are more flexible than sophisticated packages. (The program above was readily attached to the simple graph plotter in the Mathematical Association's *132 Short Programs*.)

4. Mathematics can be an exciting, dynamic, investigative science which nevertheless leads to rigorous truths.

5. The fact that an increasing sequence can be bounded above and the convergence of $(1 + 1/n)^n$ are doubtless analytic properties of the real continuum. Yet the approach above is via discrete mathematics. Or was it a numerical method? Which models which? Does the analytic formula $A = ae^{kx}$ provide a mathematical model of a discrete process, or does the bank's method of compounding interest constitute a numerical approximation to the formula? This is analogous to the interplay between discrete numerical approaches to calculus, analytic methods and numerical approximations.

Does the practising mathematician generally bother to take the limit nowadays? Would we want a formula for mortgages, for example, or would a computer algorithm do? And what of cases when no analytic method is possible?

Rate of change
My final 'case-study' illustrates my contention that short programs are as much for reading and writing as for running. This introductory lesson was given on the last day of term to a fourth-year set due to take mathematics at O-level the following term. They already had

experience of the gradient of a tangent as a rate and most of them could write simple three-line programs in BASIC. All of them had calculators. I gave them the following problem.

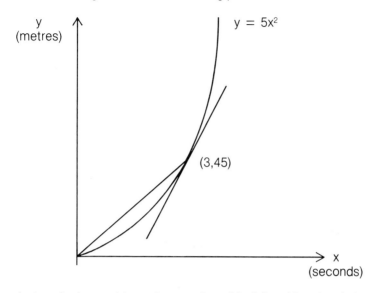

A stone is dropped from the top of a tall building. How fast is it travelling after 3 seconds if we assume that the distance fallen (y metres) after x seconds is given approximately by the formula $y = 5x^2$? We discussed how to go about calculating the gradient at the point (3,45). One pupil suggested that the answer was 45/3 = 15m/s and demonstrated graphically what he had done. I invited the class to say whether this was an over- or underestimate and why. Suitable discussion and prompting extracted the suggestion that we calculated the speed over an interval, say from 2s to 3s, and, this done, that the interval be shortened. Pupils then tabulated their results and drew conclusions.

DX	$x = 3 + DX$	$y = 5x^2$	$DY = y - 45$	SLOPE $= DY/DX$
− 3	0	0	− 45	15
− 1	2	20	− 25	25
− .5	2.5	31.25	− 13.75	27.5
− .1	2.9	. . .		
− .01				
+ .01				
+ .001			. . .	30.005

Even with a calculator this process became a little tedious. We now wanted to find the speeds at different instants; so the faster pupils were asked to write and/or complete and test the program.

```
10 INPUT DX
20 LET X = 3 + DX
30 LET Y = 5 * X^2
40 LET DY = Y − 45
50 LET SLOPE = DY/DX
60 PRINT DX,SLOPE
70 GOTO 10
```

These are just the headings from the table above.

The next step was to generalize, varying the point and then the function:

```
5  LET DX = .001
10 INPUT A
20 LET X = A + DX . . . and so on
```

By the end of the lesson some were already beginning to do this.

Conclusion
And that is quite enough for one article. I hope that I have succeeded in giving at least a glimpse of the new and exciting methods of teaching now being opened up by the computer and a hint of how it is already modifying our view of what is important in school mathematics. I did not even get started on symbolic systems; the effect of computers on mathematics, dependent disciplines and the needs of society; the use of graphic packages; other prepared software; short programs as black boxes; geometry; statistics; Cockcroft or coursework . . .

References
1. FLETCHER, T J, *Microcomputers and Mathematics in Schools: a discussion paper,* 1983 (available free from DES Publications Department, Honeypot Lane, Stanmore, Middlesex)
2. TALL, DAVID, 'Introducing algebra on the computer', *Mathematics in School,* 12, 5, 1983
3. TALL, DAVID *et al,* 'The mathematics curriculum and the micro', *Mathematics in School* 13, 4, 1984
4. HIGGO, JR, 'The microcomputer as a learning aid in sixth form mathematics, *Mathematics in School* 13, 3, 1984
5. HMI Discussion Document *Curriculum Matters 3. Mathematics from 5 to 16,* p34, HMSO, 1985

Playing algebra with the computer

*David Tall, University of Warwick, and Michael Thomas, Bablake
School, Coventry*

Ask an average 11-year-old: 'I'm thinking of a number and three
times it is six: what's the number?' and it is likely that the response
will be correct. But ask a 16-year-old to solve the problem '$3x = 6$'
and many of average ability or below will not be able to do it. The
more able students who go on to study mathematics A-level still
include a large number who make seemingly elementary slips in
algebra. So where are we going wrong?

It is clear that the abstract nature of the symbols being
manipulated is a major source of the problem. An encouraging
method of making the ideas more concrete is to introduce simple
programming in BASIC (1). To try out this idea we designed a three-
week work-schedule for a mixed-ability class of 11- or 12-year-olds,
at the top end of a middle school or first-year secondary.

As well as simple programming, the children carried out the
processes using a simple 'maths machine' made out of cardboard
with rectangular boxes labelled with letters and cardboard numbers
stored inside them. To help the children cope with different
notations used in formal algebra (such as '2x' instead of '2 $\star$ x') we
also prepared a computer programmed 'maths machine' that
accepts either version and discussed algebraic notation in parallel
with BASIC.

The full kit of 'Maths Machine', computer program, worksheets
and lesson plans was designed as a unit to be handed to a
classroom teacher ready for use (2).

To test what changes occurred in children's understanding of
algebra we gave them a little test before and after; we shall say
more about this later on. Suffice it to say at this stage that the first
class to try out the schedule was a mixed-ability group of 21 12-year-
olds. Before the schedule they made the mistakes common for children
of this age, but afterwards in a number of ways they had insights that
compared favourably with children up to three years older.

Misconceptions

Early experiences with the use of letters in mathematics can cause
serious misconceptions when children meet algebraic notation.
Often the letters are introduced as objects, so that
'3a + 5a = 8a' is interpreted as '3 apples plus 5 apples is 8
apples'. This concrete model works well to begin with, but it cannot
cope with such questions as '3a − 5a = ?'

The use of letters for units such as '100cm = 1m' also causes untold confusion in mathematics for here the letters cm and m do not stand for variables. Another common error arises from the use of letters as specific numbers such as a = 1, b = 2 in puzzles and codes. A common response given by children to the question: 'What is the total cost of x pencils at 5 pence each and y crayons at 7 pence each?' is £2.95. This mysterious answer becomes more obvious when one substitutes x = 24 and y = 25.

A child's experience of the equals sign as 'makes' in arithmetic leads them to expect that the answer to a problem must be a number, not a formula, so that 'If x = 3 and y = 5, then x + y + w = ?' poses great difficulty. One of our children tested before the algebra scheme gave the numerical answer 72. He ignored the given algebraic values altogether and used the coding w = 23, x = 24, y = 25.

The predilection for numerical answers causes a variety of mysterious responses from children that undoubtedly follow from misapplied internal strategies, such as: 'Multiply 4a by 3. Answer: 189'. 'Add 3 on to 3x. Answer: 327'. 'If I have x pieces of wood, each of length 5 metres, and another y pieces, each of length 3 metres, then what does 5x + 3y represent? Answer: 30 pieces of wood'.

Faced with such inscrutable use of algebraic notation, our aim is to use the computer to give a meaning to algebraic symbols so that children can handle them in a sensible way.

Basic ideas
If one types the statement a = 3 into a computer using BASIC followed by PRINT a + 2 then the computer prints 5. This sets up an expectation that the computer will handle algebraic notation in a sensible way. If one next types PRINT a + a then the expectation is that the response will be 6. When this response is confirmed, confidence in the computer's sensible activities is increased. If we type b = a + 1 followed by PRINT b then the expected response 4 gives further insight.

Young children respond well to the logic behind the computer's responses. To help them develop a mental picture of what is going on inside the computer, we supplemented the computer work with a physical model made out of cardboard. This is no more than two large sheets of card, one representing the screen, the other representing the variable store. The screen card is empty, the variable store card has a number of large rectangles on it, each rectangle being a store for a number (Figure 14). This 'maths machine' — named after the work of L Booth (3) — is completed by a number of pieces of card marked with letters and numbers.

Instead of the children's merely pressing keys and getting instant answers, this maths machine required them to perform their own calculations and to be involved in the model, physically transferring answers from one part of the machine to another.

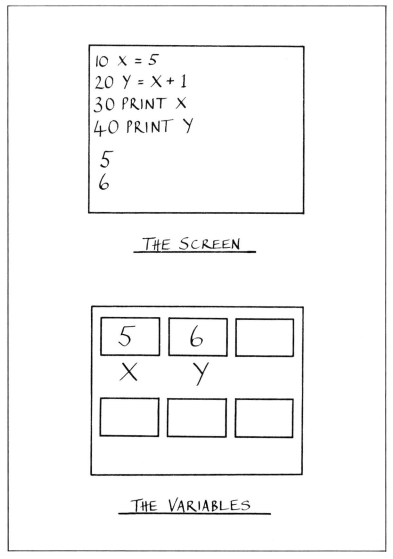

Figure 14. The cardboard maths machine

The boxes are initially without labels, but the assignment a = 3 is represented by labelling the box 'a' and putting a piece of cardboard inside containing the number 3. If a further assignment is made, b = a + 1, then it is necessary to label another box 'b', look into box 'a', note the number 3 inside, add one, making 4 and putting 4 inside the 'b' box. At the end of this activity there are now two boxes, 'a' containing 3 and 'b' containing 4.

These techniques can be used not only to calculate numerical values, but also to compare different ways of carrying out the calculation. For example, one may check that a + a and 2 ∗ a always give the same result, or 2 ∗ (a + b) gives the same result as 2 ∗ a + 2 ∗ b. An element of competition can be introduced with a group of children operating a computer to calculate 2 ∗ (a + b), comparing their results with a group using a physical maths machine to calculate 2 ∗ a + 2 ∗ b.

Thus the variables and their current values were constantly before the eyes of the children. Acting out the workings of the computer in this way, seeing the 'inside of the computer' as one put it, was a big hit with them, even though they found they had to work a bit harder', as one of them said, 'you could see what was happening but you had to do all the work'. The pupils were rotated in groups of three between the computers and the maths machines so that all had ample experience with each.

Another tool designed specifically for this work was a computer program which would enable problem-solving in algebra at both simple and more advanced levels. The program, which allowed normal algebraic input for the functions (Figure 15), was designed to enable children to develop their algorithmic thinking through tackling structured practical algebraic problems by a concrete process, at the same time reinforcing and using the mental visual model previously encouraged. In order to allow comparison of expressions side by side, the program displays two boxes with their corresponding formulae written over the top, the boxes containing the values of the formulae. Thus there might be boxes on the screen in which one could input values for a and b, and two others in which the values of 2 ∗ (a + b) and 2 ∗ a + 2 ∗ b were calculated and displayed. The advantage of this programmed maths machine is that the input of formulae can be designed to accept ordinary algebra as well as BASIC notation, thus giving a concrete meaning to standard algebraic symbolism. The display designed for use in the work-schedule allows some variables to be input and others (representing constants) to be fixed (Figure 16). This program proved to be extremely popular with the children.

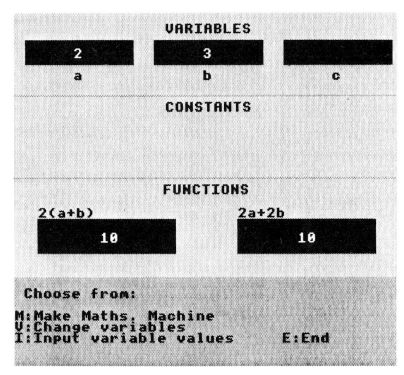

Figure 15. Screen display from algebraic problem-solving program showing variables and functions input

Since the end of the teaching programme, a second computer program, an 'algebraic calculator', suitable for all ages in the secondary school and also employing algebraic input, has been developed. The calculator displays three columns, one for the label of the variable, one for its current numerical value and a third for an optional formula which is used to calculate the value of the variable in terms of other variables (Figure 17).

Resources
To teach algebra through 'hands-on' computing clearly requires access to a number of computers. But by using a combination of computers, cardboard maths machines and paper-and-pencil calculations we found it practicable to cope in a class of 21 children using only three BBC computers (the full resource of the school at the time). The best plan is to have working groups of three or four

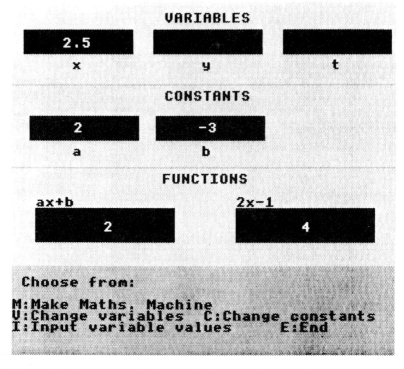

Figure 16. Screen display from algebraic problem-solving program showing variables, constants and functions input

children to each computer. Cooperation produces discussion and is often more helpful than solitary effort.

The activities

The trial run of the algebra schedule was carried out at Woodloes Middle School, Warwick, in a class run by Mrs Elizabeth Stanley. The first activity was to familiarize the children with elementary aspects of the computer. This was done as a class lesson, sitting around a computer with different children taking it in turns to type things on the keyboard. [It is useful to double the height of the printing on the screen so that children can see it from a distance. This can be done on the BBC computer using a program on the Mathematical Association disc of *132 Short Programs* (4).]

As the class teacher in this case was also the English teacher, it was natural to start off with print statements such as PRINT "Hello there"

store	value	formula
x	→1	
a	5	3*x+2
b	5	3x+2
c	9	3(x+2)
d	9	3x+6

Figure 17. Screen display from algebraic calculator

(Press RETURN) using speech marks. After taking it in turns to type in a few such statements, the children then investigated what happens when the speech marks are left out in 'PRINT good idea' and were greeted with 'No such variable'. Several more attempts missing out the speech marks led to the same error message. At this stage the teacher went to the blackboard and explained how the computer sets up stores for numbers. The store must have a label or name. When we type PRINT A without speech marks, the computer searches its memory for a store labelled A so that it can print out the number stored there. A store which can contain any number is called a 'variable'. If the computer cannot find a store named A, it prints the 'No such variable' message.

Making variables

The children were asked how they might get the computer to label a store A and put a value, say 6, into the store. Not long ago few would have known what to do, but now every class is likely to have the computer 'expert' who suggests typing A = 6. Other children can

65

take it in turns to suggest labels and print the values. They are likely to amuse themselves with all sorts of names: WHAM = 5. PRINT WHAM gives the value 5, but THE BEATLES = 4 yields 'Mistake' because a variable must be a single word. It is a good idea to allow such mistakes to be made for it soon leads to an idea of the allowable symbols for a variable. At the same time the children become familiar with essential features such as using the RETURN key at the end of an input and the SHIFT key to enter the = sign. It is also useful to use both upper- and lower-case letters for variable names.

Cardboard maths machines

Before doing much work with the computer it is beneficial to introduce the children to the cardboard maths machines which allow them to be actively involved. Several of these were kept at the side of the classroom and more could easily be made by the children themselves. There was a plentiful supply of letters for names and numbers, all written in large print on pieces of cardboard. It is possible to put the 'screen' on one side of the room and the variables on the other to allow competitive games.

The analogy between the computer and the maths machine was emphasized. When the statement A = 6 is typed into a computer it checks to see if it has a store labelled A. If it does, it simply puts the number 6 in the store. If not, it labels a new store A and puts the number 6 into that. The cardboard maths machine is used in the same way. To operate it different children are given individual jobs. One can be the human operator, issuing instructions by placing them on the screen, either as direct commands or as part of a program. The others share the internal computer jobs. These involve carrying messages from the screen, looking after the labels, inserting number cards into the stores and performing arithmetic operations where required. If, for example, the operator issues the command A = 6, the message is taken to the labeller who looks for a box marked A; if there isn't one, he labels a new box, and a card with the number 6 on is put into it. Faced with the statement PRINT A it is necessary to search for a box marked A to check the number inside and place another card with the same number on the 'screen'. If there is no box labelled A, the 'computer' can say 'no such variable'!

Algebraic investigations

Once the children have an idea how the computer and the maths machine work they can be set to do some investigations. By setting up a rota system children working in groups of three or four can alternate between the systems. A typical investigation might be to consider the effect of:

X = 5
Y = X + 1
PRINT X
PRINT Y

A discussion is necessary for those on the maths machine to understand what to do. The second line involves them looking in the box marked X, adding 1, marking a new box with the label Y and putting the value calculated into it.

Exercises can be done competitively. Do the computers agree with the machines? Other sequences such as

A = 7
B = A − 3
PRINT A
PRINT B

can be tried in the same way.

At a later stage the signs * and / were introduced for multiplication and division and further investigations were carried out using these operations.

Short programs

The idea of a program as a list of instructions was introduced in a later lesson. On the computers, groups typed the programs

10 X = 3	10 X = 3
20 Y = 5	20 Y = 5
30 Z = X + Y	30 Z = Y + X
40 PRINT Z	40 PRINT Z

and then issued the command RUN.

Do these programs give the same answers? Why? Change the values of X,Y. Are the answers still the same? In the process children learn how to type programs and to modify them by retyping the line that requires changing.

Longer programs can be studied with three variables A,B,C being specified and a dependent variable D given either as D = A + B + C or D = B + A + C being calculated and printed. Or programs specifying X and printing Y given by Y = 2 * X or Y = X * 2 or Y = X + 2 can be compared to see when they give the same values.

On the maths machine a program is RUN by having a list of instructions placed on the screen of the machine and obeying them one after the other. Further investigations can be carried out along these lines on the real computer or the maths machine.

The use of brackets
How could a computer be told to add 1 to x and then multiply the result by 3? Discussion with the children produced the expected x + 1 for the sum, then 3 * x + 1 for the final answer. It was tried in a program

10 x = 6
20 y = 3 * x + 1
30 PRINT y

What should the answer be? Certainly x + 1 should be 7 and 3 times 7 is 21. What happens when the program is run? It doesn't work!

This is the time to discuss how the computer interprets the notation and to get the children to suggest how we might write the sum so the answer is that expected. Brackets are introduced, with line 20 modified by writing 3 * (x + 1). This time it works. So the children see that 3 * x + 1 and 3 * (x + 1) are not the same.

They can investigate the results of printing 3 * (x + 1) and 3 * x + 3. What happens?

Further investigations using the computer and the maths machine illustrate the use of brackets.

INPUT
The INPUT command was introduced in the form

10 INPUT x

so that when RUN, the computer puts a question mark on the screen and waits for a value for x to be typed in. After a number is typed (followed by RETURN) the computer stores the number in the store x.

The GOTO instruction was introduced to allow programs such as

10 INPUT A
20 B = 2 * A
30 PRINT B
40 GOTO 10

to be used and to compare the results of

10 INPUT X	10 INPUT X
20 Y = X + X	20 Y = X * 2
30 PRINT Y	30 PRINT Y
40 GOTO 10	40 GOTO 10

68

Are X + X and X * 2 the same? What about 2 * X? Modify the program again to see.

Applications
The ideas were used to solve practical programs. For example:

If we know the length and width of a rectangle, how can we calculate the area? Can we use the computer or maths machine to help us do this for any rectangle?

The idea is to move towards a program of the kind:

```
10 INPUT L
20 INPUT W
30 A = L * W
40 PRINT A
50 GOTO 10
```

A second example is to suppose that the computer has a store S in which is stored the number of sweets we wish to buy at five pence each — how can we tell the computer to print the cost?
(PRINT 5 * S.) This leads to an extension, namely, printing out the cost of S sweets at five pence each plus C comics at 12 pence each by printing 5 * S + 12 * C. The children were then able to do a sheet of examples of similar applications, working with computer and maths machine once more.

The link with algebra
It is important to know that standard algebraic notation used at the moment misses out the multiplication sign and writes 2x instead of 2 * x and 3 (x + 1) instead of 3 * (x + 1). Throughout the lessons allusions were made to this notation and the children did paper-and-pencil exercises to translate algebraic notation to BASIC and *vice versa*.

The computer maths machine
The children were introduced to the computer maths machine mentioned earlier. They were able to specify letters as labels for variable stores and to type in one or two functions expressions. For instance, with a single variable x, they could attempt to answer 'when does 5 + x = 9?' by typing in the expression 5 + x and calculating its value for various values of x.

The same method could be attempted for the slightly harder expression 'when does 3 − x = 2 + x?' by typing in the two expressions 3 − x and 2 + x and calculating the expressions for various values of x. Even inequalities can be attempted: 'when is y − 2 > 5?' by typing in the expression y − 2 with variable y and finding values of y that make y − 2 bigger than 5.

Furthermore expressions such as 2x and 2 ∗ x or x + x and x 2 can be compared and seen to be the same for all values of x (Figure 18). Or the use of brackets can be investigated to compare 3 (x + 2) and 3x + 2 or 3 (x + 2) and 3x + 6.

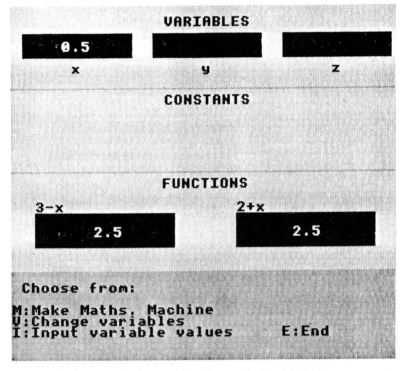

Figure 18. Screen display used to show relationship between functions

The benefit of using the computer maths machine is that it continues the 'picture' of the variable concept previously used, but now the children can see the values of the variables as they vary. The 'varying' becomes part of their concrete experience. Furthermore they are using this 'variability' to tackle problems that are in a sense 'real' to them and provide a mental picture to use in the future.

Comments from the pupils and teachers

It seems insufficient somehow to say that the pupils enjoyed this work, but they did, in fact they enthused over it. It is true that they had not previously been exposed to work on the computer, but as many will know, this enthusiasm wanes slowly, and the enthusiastic comments of the children on interview were directed at the content of the work as well as the computer.

'It was good fun' said one, 'we have never done anything like it before.' 'It was some of the best work this year', agreed another. 'I liked the challenge of finding unusual answers.'

The classteacher's comments were equally positive.

'This was a very worthwhile project which proved to be very pupil-orientated. It is enjoyable, interesting and thought/discussion-provoking between pupils and between pupils and the teacher. All pupils enjoyed working with the maths machine as it helped most realize what was happening "behind the screen". They also felt in total control of what they were doing. They were also a good aid for mental arithmetic. My feeling is that they brought life and activity to the mathematics session. This work is a very useful way of introducing children to the workings of the computer. It is simple and valid . . . The need to give clear commands aided logical thinking. On the whole the worksheets were very good and varied in content. They were very well liked by the pupils. Overall the pupils were eager from the outset and this did not wane in any way right through to the last session.'

This enthusiasm for the project also carried over to the other year-teachers in the school, who expressed their desire to use the work in the future.

What did the children learn?

After the schedule of work the children's views on algebra had changed markedly. Typical comments were that they 'didn't know letters could be used this way', and 'I will find basic maths easier now'. Was this borne out by the results? The main results of the teaching experiment certainly showed that those involved in the work had a significantly better understanding of algebra at the end. Some specific examples of individual improvements in understanding shown by the pre- and post-tests were as follows.

(a) Individual changes

1. Before:

When does L + M + N = L + P + N?
Always, never, sometimes when . . .

Answer — never

After:

When does M + P + N = N + M + R?
Always, never, sometimes when . . .

Answer — sometimes when R = P

2. Before:

For what value or values of a is a + 5 > 8?

Answer — 2

After:

For what value or values of a is a + 3 > 7?

Answer — 5 and over

3. Before:

If x = 3 and y = 5 then x + y + w = ?

Answer — 72

After:

If x = 2 and y = 6 then x + y + w = ?

Answer — 8 + w

4. Before:

A figure has n sides each of length 2 centimetres. What is the perimeter of the shape?

Answer — 8

After:

A figure has p sides each of length 3 centimetres. What is the perimeter of the shape?

Answer — 3p

(b) A comparison with the CSMS results

The following is a comparison of the results of the experimental group with the average facilities of different age-groups as given by the results of the Concepts in Secondary Mathematics and Science (CSMS) algebra study (5), on some specific questions taken from the tests.

1. When does $M + P + N = N + M + R$? Always, never, sometimes when . . .

Pre-test	0%
Post-test	42.9%
CSMS 15-year-olds	27%

2. Write down the area of a rectangle whose length is X centimetres and whose width is W centimetres.

Pre-test	4.8%
Post-test	71.4%
CSMS 14-year-olds	68%

3. If $x = 4$ and $y = x + 3$ then $y = ?$

Pre-test	(not given)
Post-test	81%
CSMS 15-year-olds	70%

4. A figure has p sides each of length 3 centimetres. What is the perimeter of the figure?

Pre-test	0%
Post-test	42.9%
CSMS 15-year-olds	41%

5. If $x + y = 5$ then $x + y + z = ?$

Pre-test	4.8%
Post-test	28.6%
CSMS 13-year-olds	25%

These are extremely favourable comparisons, considering that the average age of the experimental group was 12 years 3 months, and that they had not had any formal instruction in algebraic techniques. Who knows what they may achieve when they have done some 'real' algebra!

Where do we go from here?

It is clear from our experiment that practical algebra on the computer can be attacked earlier and more successfully than is the case with traditional paper-and-pencil manipulation of symbols. The experiences we had here can easily be replicated in other classes provided that the teacher has a flexible approach and is willing to cope with the variety of responses that is likely to occur from children's investigations. The lessons are very much in line with the Cockcroft recommendations, including all aspects of:

— exposition by the teacher
— discussion between teacher and pupils and between pupils themselves
— appropriate practical work
— consolidation and practice of fundamental skills
— problem-solving
— investigational work.

The techniques are still very much concerned with simple notation rather than algebraic manipulation such as simplifying, factorizing, solving equations and so on. This we still have to design and try out. But it is hoped that the strength of the insights the children have developed will stand them in good stead. The horrors in algebra that occur later in the school, even in the sixth form, will not be solved by revamping the curriculum at this later stage. They must be attacked at the outset when the concepts are first introduced so that the children have a meaningful mental picture of what the symbols represent.

References

1. TALL, D O, 'Introducing algebra on the computer'. Mathematics in Schools 12, pp37-40, 1983
2. THOMAS, M O J, The Effects of BASIC Computer Programming on the Understanding of the Use of Letters as Variables in Algebra, MSc thesis, Warwick University, 1985
3. BOOTH, L R, Misconceptions Leading to Error in Elementary Algebra (Generalized Arithmetic), PhD thesis, Chelsea College, London, 1983
4. THE MATHEMATICAL ASSOCIATION, 132 Short Programs for the Mathematics Classroom, Mathematical Association, Leicester, 1985
5. KUCHEMANN, D E, in HART, K M (ed), Children's Understanding of Mathematics: 11 – 16, John Murray, 1981

The use of the micro in probabilistic simulations

F R Watson, Department of Education, University of Keele

Probability and statistics are a part of school mathematics — 25 years ago they were not so, but, unlike some other topics which formed part of the modern mathematics innovations of the 1960s, they have won acceptance, perhaps as a recognition that so much of our life and the world around us involves probability. The micro is a powerful aid to work on probability, and I want to indicate some of the possibilities in the field of simulation.

1. Random numbers

All micros, and some handheld calculators, have a random number generator (strictly, a pseudo-random number generator); the idea of randomness is a difficult one, both philosophically for the expert and conceptually for the beginner. Its essence lies in the fact that we can predict, with reasonable accuracy, overall behaviour — the global picture, or events 'in the long run', but we are totally incapable of predicting in detail, ie, 'what will happen next time'. Some illustrations may help. A dart is thrown by a novice from a fair distance — will it fall in the upper or lower half of the dartboard? To left or right? Near the centre or near the edge? (The answer in the last case may be nearer the centre — though that will depend on how much of a novice our darts player is!) Another example — try throwing a small stone on to a pebbly beach (*not* near other people! My 'record' is six bounces). The behaviour of the bouncing stone is quite unpredictable.

We can use the computer to illustrate randomness too: Program 1 (for the BBC) displays, when the space bar is pressed, a small square at random positions on the screen, in a random colour; the colour number (from 1 to 7) is displayed at the top of the screen. Pupils can be invited to guess:

— where the next square will appear
— what colour it will be
— what number will appear next.

Program 1
```
 10 MODE2
 20 FOR I = 1 TO 100
 30    C% = RND(7):PRINT;C%;
```

```
 40    PROCSQ (C%)
 50    Z = GET
 60    NEXT I
 70 END
 80
 90 DEF PROCSQ (C)
100 GCOL 0,C
110 MOVE RND(250)*4,RND(200)*4
120 PLOT 0,16,0
130 PLOT 81,0,16
140 PLOT 0,−16,0
150 PLOT 81,0,−16
160 ENDPROC
```

If the pupils are able to say, 'We just do not know where the next square will fall, or what its colour will be, but we predict that about 50 per cent of the squares will be in the top half of the screen, and about 14 per cent will be 'red', then they have some understanding of this 'random' behaviour.

Program 2 provides an analogue of the dart-throwing experiment. A moving point draws a horizontal line at the bottom of the screen, first in yellow, then in red, then in yellow . . . When you press the space bar, a black spot is placed on the line; at the same time, a random number in the range 0 − 1 is printed, corresponding to the position of the moving point at that instant. Pupils can see that the point moves uniformly, constantly retracing its path from 0 to 1, and that the random number printed records its position when the space bar is pressed. (They can practise estimation of the position, or measure reaction times, too — but that is another story.)

Program 2
```
 10 MODE1
 20 @ % = &01020503
 30 C% = 1:FF = − 1
 40 MOVE 0,200:VDU5:PRINT"O":MOVE 1020,200:PRINT"1":VDU4
 50 REPEAT
 60    MOVE 0,100
 70    FOR I = 0 TO 128
 80       DRAW 8*I,100
 90       GCOL 0,C%
100       Z = INKEY(2)
110       IF Z = 32 THEN PRINTI/128:GCOL 0,0
120       IF Z = 13 THEN FF = − 1 + ABS(FF)
130    NEXT I
```

```
140    IF FF THEN C% = 3 – C%
150    GCOL 0,C%
160    UNTIL FALSE
170 END
```

(The @ instruction in line 20 is merely to control the format of output.)

By pressing RETURN at any stage, the red-yellow-red colour change is inhibited (line 120 changes 0 to – 1 and – 1 to 0), so that it is no longer possible to see where the moving point is; after some time it could be anywhere — and, indeed, pressing the space bar will produce a random number in (0,1) — and perhaps some surprises! (A second press of RETURN restores the original situation.)

Clearly, given perfect knowledge, we could state exactly where the point was at any instant — but in the absence of omniscience, we have to treat its position *as if* it were random.

This is an important aspect of the use of randomness in simulating events which, in fact, are quite determined — but in ways we cannot unravel.

Random patterns of lines, random shapes, random walks illustrating 'Brownian motion' (Mathematical Association, 1985), the random choice of numbers appearing in games such as 'Blocks' (A Straker, 1984) provide other opportunities for illustration.

2. The idea of simulation
It seems best to approach this idea, also, in a very 'concrete' way; I have found simulation of two-way traffic on a busy street to be a good introductory situation (GREENX; Watson, 1980, pp62, 74).

A girl waits to cross the road — she will not cross if there is a car within 100 metres coming towards her. The arrival of a car in a period of 10 seconds can be simulated by tossing a coin or rolling a die ('car coming' = H, or 'car coming' = 1 or 2 or 3). What is the probability that she will be able to cross?

This simple situation can be used to illustrate some basic ideas — increase the traffic flow (rush-hour) by taking 'car coming' = 1 or 2 or 3 or 4 or 5; experimental verification of the multiplication law: $p(A \text{ and } B) = p(A) \star p(B)$. (For example, suppose $p(L) = prob(car from left) = 1/3$ and $p(R) = 1/2$; then $p(no car) = 2/3 \star 1/2 = 1/3$ = probability of successful crossing, so that about 33 per cent of attempted crossings should be successful.) What if we had a zebra crossing, or pedestrian-controlled traffic signals?

It seems important *not* to use the computer in the first stages — a jumble of numbers on the screen can be completely unintelligible if their relationship to the underlying situation has not previously been established. Sometimes it is necessary to take a very simple approach. I have clear memories of a lesson with second-years where a pencil, paper and blackboard representation of this situation proved too difficult — I *ought* to have used bricks, or biros or books on the front desk to represent cars and children crossing, and, in similar situation, have subsequently done so! When the manipulation of objects representing cars and people is seen as superfluous, we can begin to record the situation with pencil and paper — and when the die rolling and the recording become too slow and tedious, it is natural to ask, can we use the computer to help? (See Program 3).

Program 3

```
 10 REM CROSSING THE ROAD
 20 R% = RND( − TIME):REM ★★ RANDOMIZE ★★
 30 PRINT "INPUT PROB OF VEHICLE      ARRIVING FROM
L"
 40 INPUT L
 50 PRINT "INPUT PROB OF VEHICLE      ARRIVING FROM
R"
 60 INPUT R
 70 PRINT "INPUT NUMBER OF TIME PERIODS"
 80 INPUT T
 90 N = 0
100 Y = 0
110 FOR I = 1 TO T
120    IF RND(1) > L THEN 150
130    N = N + 1
140    GOTO 190
150    IF RND(1) > R THEN 180
160    N = N + 1
170    GOTO 190
180    Y = Y + 1
190    NEXT I
200 PRINT
210 PRINT
220 PRINT "NUMBER OF TIME PERIODS ";T
230 PRINT
240 PRINT "Y = ";Y; "N = ";N; "Y/T = ";Y/T
250 PRINT
260 PRINT "FOR PROBABILITIES L = ";L; "R = ";R
270 END
```

3. Mimicking dice and coins

Probability lends itself well to an experimental approach. Indeed, there is something phoney about not comparing the predictions of a theoretical model with the outcomes of experiment. (Many years ago a teacher remarked after a discussion of group results on coin-tossing (in which the usual quota of 'freak' results cropped up): 'I would not dare to do this with my girls — it would destroy their faith in probability theory'.)

However, once the structure of an experiment is clear, recording 200 tosses of a coin or rolls of a die becomes tedious and noisy, and there is a case for moving to computer simulation after some of the simple experiments have been tried manually to display the ideas and provide the link with 'reality'.

Judicious multiple use of appropriately recorded events can ensure we get full value out of our experiments; the sequence HHTHTTTHHT . . . not only tells us, say, that 100 tosses produced 53H and 47T, but can also be used to investigate tossing two coins (by pairing), triples, sets of four, etc, whether H is more frequently followed by another H or by T ('The Australians have won the toss the last three times, so the odds are on England'), the distribution of lengths of runs, etc.

At a more sophisticated level, all these experiments referred to, and many more, can be simulated by computer, often by a series of adaptations of essentially the same simple program (see Mathematical Association, 1985). Here, though, I shall refer only to some simple considerations worth noting.

On the BBC micro, the statement $X = RND(6)$ produces a random integer chosen from (1,2 . . .6) and assigns it to X; similar commands are available on the other micros currently used in schools. This makes it easy to mimic the behaviour of dice and coins and to produce random digits (by $RND(10) - 1$); however, it is also instructive to consider how random numbers in the range $0 = < R = < 1$ may be used for this purpose. Gillian Oakes (1984) describes a delightful lesson in which the pupils gradually sorted out for themselves how to use this to mimic throws of a die; they had aready heard of INT from their teacher, and she prompted them, via $X = 6 \star RND(1)$ and $6 \star RND + 1$ to eventual success. It would have been so easy — and so much less productive in the long term — to have said 'To solve this problem we use $X = INT(6 \star RND(1)) + 1$'.

4. Examples of situations for simulation

Several situations may be used — the following will no doubt be familiar. A simple popular one is 'radioactive decay'; the work may be based around a program such as:

Program 4
```
10   INPUT N:S = N
20   REPEAT
30      FOR I = 1 TO N
40         IF RND(1) < 0.2 THEN S = S − 1
50         NEXT I
60      N = S: PRINT S
70      UNTIL S = 0
80   END
```

Alternatively, many physics packages contain graphics versions on this theme (eg, an Apple program, distributed some years ago by MUSE, which displays the 'atoms' as a rectangular array — they change colour as they decay').

Pupils may plot a graph showing the progress of the simulation — or we can get the computer to plot it for us. We note that in Program 4, S 'ought' to be reduced by 20 per cent at each stage, and the theoretical and simulated results may be compared. (This situation corresponds to the decay of a substance with a 'half-life' of 3.1 years — ie, the amount remaining will be reduced to half of that initially present after 3.1 years, to one quarter after 6.2 years, and so on; we need to solve the equation $(4/5)^n = 1/2$, which leads to $n = 3.106$).

Another simple situation is given in ICL/CES (1979, p254); we are planning a camping holiday and will abandon it and come home if we get rain on five consecutive days; if the chance of any given day being wet is 0.3, how likely are we to abandon our holiday?

Program 5
```
 10  REM III254 WEATHER
 20  REM VALUES SET IN LINE 40
 30  REM SPEED CAN BE VS very slow (day by day), S slow (5 day
groups), Q quick (5 day totals) or VQ very quick (freq table only)
 40  N = 5: P = 0.7:TRIALS = 40
 50     INPUT "SPEED: VS,S,Q, or VQ",A$
 60  DIM D(N)
 70  FOR J = 0 TO N:D(J) = 0:NEXT J
 80  FOR Z = 1 TO TRIALS
 90     F = 0
100     FOR I = 1 TO N
110        X = RND(1)
120        IF A$ = "Q" OR A$ = "VQ"THEN GOTO 150
130        IF X < P THEN PRINT ; "F"; ELSE PRINT;"W";
140        IF A$ = "VS" THEN G = GET
```

```
150      IF X < P THEN F = F + 1: REM FINE
160      NEXT I
170      IF A$ = "S" OR A$ = "VS" THEN PRINT ;" ";F; "fine days":
G = GET:PRINT
180      IF A$ = "Q" THEN PRINT ;F; " ";
190      D(F) = D(F) + 1
200      NEXT Z
210 PRINT"No. of fine days"
220 PRINT"  0   1   2   3   4   5"
230 PRINT
240 PRINT "Frequencies"
250 @ % = &00000005
260 FOR J = 0 TO N:PRINT D(J);:NEXT J
270 PRINT
280 Q = 1 − P
290 D(0) = Q ^ N:R = P/Q
300 FOR J = 1 TO N
310      D(J) = D(J − 1) *R *(N − J + 1)/J
320      NEXT J
330 PRINT:PRINT"Theoretical values":PRINT
340 PRINT"  ";
350 @ % = &00020105
360 FOR J = 0 TO N :PRINT D(J)*TRIALS; :NEXT J
370 PRINT
380 @ % = 10
```

Here four versions of output may be obtained, showing different
degrees of detail, ranging from a day-by-day report of weather
(choice VS), to a frequency table of number of fine days in a group of
five for 40 trials (choice VQ) — see line 30.

Traffic situations provide plenty of material for simulations — the
pedestrian crossing problem of Program 3 can be modified — say
by incorporating pedestrian-controlled traffic lights — how much
delay does this cause to cars? One-way traffic caused by road
repairs can be investigated (ROADUP) — how should the timing
sequences of traffic lights be adjusted to cater for different rates of
traffic flow? Similar work can follow on T-junctions and crossroads
— would traffic lights reduce the aggregate misery, measured in
motorist-impatience units (defined as 1 miu = 1 driver waiting for a
period of 1 second)? Further discussion of some of these situations,
with related programs, may be found in Watson (1980), which also
gives other suggestions for simulation.

As a final example consider another real-life situation, that of
stock control. ICL/CES (1979, p247) uses an illustration based on a

furniture warehouse — you may prefer to consider keeping the school tuckshop or canteen fully stocked. Program 6 simulates the arrival of orders of various items of furniture. It is a micro version of an activity suggested in the text, using dice to generate random numbers of desks ordered 0,1, . . . ,5 (ORD). Three new desks are made each day, and added to the stock that evening (STKE), unless the existing stock that morning (STKM) was more than eight desks, when none are made (line 80).

Program 6
```
 10  REM IIIP247 SIMULATION
 20  @ % = &00000405
 30  PRINT'' DAY  STKM ORD  STKR ADD  STKE''
 40  STKM = 6:OOS = 0
 50  FOR I =  1 TO 20
 60     NEWORD = RND (6)
 70     IF NEWORD = 6 THEN NEWORD = 0
 80     IF STKM < 9 THEN ADD = 3 ELSE ADD = 0
 90     STKR = STKM − NEWORD
100     STKE = STKR + ADD
110     PRINT I,STKM,NEWORD,STKR,ADD,STKE
120     IF STKR < 0 THEN OOS = OOS + 1
130     STKM = STKE
140     NEXT I
150  PRINT OOS;''  = no. of times out of stock''
160  @ % = 10
170  END
```

5 The interplay of theory and simulation results
It is valuable to compare simulation results with predictions from theory where these are available; this can serve to support both aspects, illustrating the mathematics (as in Section 2 above), and showing the power of a prediction, which can short-cut a lot of simulation of experiment — as well as the limitations of theoretical predictions, especially in some complicated situations. The mathematics involved can rapidly become too difficult for classroom discussion, but some results may be quoted and investigated (such as the $n^{1/2}$ law in random walk problems), and others will not be beyond the capacity of A-level students who can *use* their maths to explain the behaviour of a simulation; examples might be absorption in a linear random walk, (Watson, 1980, p121), or two-stage radioactive decay (Watson, 1980, p118).

Common-sense suggestions can also be checked — it is reasonable to predict in the ROADUP simulation of Section 4 above, that priority should be allocated in proportion to the traffic density in the two directions, and that if traffic is heavy, changes of priority should be infrequent.

It is worth making the point that simulation provides the only practicable way of investigating some situations; for example, though queueing problems have been intensively studied by mathematicians, situations involving two or more linked queues are almost completely intractable in analytic terms. (Another point of interest is that instead of using a mathematical analysis to predict the result of a simulation, we sometimes find it convenient to reverse the process, using simulation to obtain approximations to equations which are too complicated to be solved by any other method.)

6 Real-world uses of simulation

The real-life applications of simulation should be mentioned to pupils — planning urban development, designing road systems, power stations, and oil refineries, predicting population growth and economic development, are some of the areas in which simulation is used. Of course, planning ahead does not always guarantee success, but there is no harm in a little forethought! Better to consider the probable effect of installing traffic lights or a roundabout before holes are dug or concrete is poured. Ormell (1979) provides a nice illustration: simulation of a proposal for operating a minibus service in Adelaide indicated it would not be profitable; a businessman who thought he knew better, abandoned the project after one day's operation!

References

ICL/CES, *Computer Studies, Book III,* ICL/CES, Reading, 1979
MATHEMATICAL ASSOCIATION, *Short Programs in the Mathematics Classroom*, Mathematical Association, Leicester, 1985
OAKES, G, *Teaching Mathematics with a Microcomputer,* MEd dissertation, University of Keele, 1984
ORMELL, CP, *The Electronic Bus-Stop,* Mathematics Applicable Group, University of Reading, 1979
STRAKER, A, *MEP Primary Mathematics Software Pack,* Microelectronics Education Programme, 1984
WATSON, FR, *A Simple Introduction to Simulation,* KMEP, University of Keele, 1980

The use of microcomputers to support mathematical investigations

Derek Ball, Leicester School of Education

Introduction
Computers are increasingly being used to support the teaching of mathematics, and many teachers and educationalists now see the use of computers as a way of introducing new styles of teaching and learning. Teachers have not, on the whole, found it easy to implement the advice of Paragraph 243 of the Cockcroft Report (1) and to include discussion, problem-solving and investigation into their teaching of mathematics. But computers appear to encourage pupils to discuss mathematics and also seem able to provide situations which lead them to solve problems and investigate mathematics for themselves. This article considers some of the ways in which computers can be used to support investigations.

Programs that control an investigation
Software writers, aware of the increasing emphasis being placed on mathematical investigations, have not been slow to produce investigation programs. Some of these appear designed to provide all that is necessary for introducing and completing an investigation within a single program.

Programs designed to help pupils investigate the 'Frogs' problem provide a good example. 'Frogs' is a pegboard problem which was often introduced to pupils by more adventurous teachers as a 'people game'. Seven chairs are lined up at the front of the classroom. Three girls sit on the three chairs at the left-hand end, and three boys on the three chairs at the right-hand end.

| G | G | G | | B | B | B |

Figure 19. Layout for acting out the 'Frogs' problem

Boys and girls may then move either by sliding on to an adjacent vacant chair or by 'hopping' over one other person on to a vacant chair. The object is for the girls and boys to change places. This situation can be investigated by posing and answering questions. Here are a few such questions

(a) How many moves did thay take?
(b) Could they have done it in fewer moves?

84

(c) Was it easy to count the moves?

(d) On which moves did they 'hop'?

(e) How many moves would they need if there were only two girls and two boys?

(f) What happens if the number of girls is not equal to the number of boys?

(g) What patterns do you get if the number of boys and girls are kept equal, but that number is varied?

(h) What pattern do you get if there are always two more girls than boys?

(i) What if there are 60 girls and 60 boys?

(j) What if there are two spare chairs in the middle?

(k) What if you have to 'hop' over two people instead of one?

Teachers may not always want to investigate 'Frogs' by using people: they may decide that the people concerned would behave badly, or they may want individuals or small groups to investigate the problem at some length. The problem can obviously be investigated using pegboard or counters or any objects that can be conveniently moved around.

There are several reasons why teachers may want to use one of the 'Frogs' computer programs to help pupils investigate the problem. Firstly, the programs are often attractively produced, with good graphics and amusing sounds, and may encourage pupils to take an interest in the investigation. Secondly, they may make it easy for teachers to encourage small groups of pupils to investigate the problem. (Such small-group work is often considered desirable in mathematics learning but many teachers find it very difficult to manage successfully.) Also use of the programs may make the initial stages of the investigation easy for pupils: the programs may ensure that pupils follow the rules of the problem properly, they may remove from pupils the need to count the moves, and so on.

Removing the need to count the moves may, alternatively, be considered to be one of the disadvantages of all the 'Frogs' programs currently on the market. When pupils investigate mathematics they are gaining the opportunity to be mathematicians, researching into their own problems. Mathematicians researching situations such as the 'Frogs' problem need to be systematic, they need to be careful and accurate, and they need to be able to spot their own errors. If pupils are using programs which force them to be systematic, which take away the need to be careful and which point their errors out to them, thay do not themselves develop these qualities.

85

Another disadvantage of the 'Frogs' programs available is that none of them allows many of the questions listed above to be investigated; in other words, the programs are only of any use if pupils want to investigate those problems considered interesting by the program's designer. The Cockcroft report suggests that one of the main benefits to be gained from investigating is encouraging pupils to ask 'what would happen if?' It is precisely this question that they are unable to ask when they use the 'Frogs' programs.

Another disadvantage of some of the 'Frogs' programs is that they are concerned at too early a stage with answers. One program, for example, asks pupils how many moves they think are required long before they are in a position to know. Other programs present answers already obtained in the form of a table displayed on the screen. This may be useful (although recording systematically on paper the data you have obtained is an important skill to learn) but the information is sometimes presented in a way intended to provide clues for those using the program. This has two disadvantages: it prevents pupils from having the satisfaction of trying to make sense of patterns of numbers without help; and it once again focuses attention on those aspects of the patterns that the program designer thinks are interesting and effectively prevents users from considering other aspects.

A fuller discussion of the 'Frogs' problem and of the advantages and disadvantages of particular programs is given in *Micromath* 1, 2 (2).

Other programs that provide different investigations for pupils may have similar advantages and, more particularly, disadvantages. A program is likely to have the same advantages and disadvantages as the 'Frogs' programs described in this section if it poses the problem for students. In this case, the chances are that the program forces students to tackle only certain aspects of the problem and to tackle those aspects in ways considered appropriate by the designer of the program.

Programs that set the scene for an investigation
Some computer programs provide mathematical games for pupils to play. Others provide tools for pupils to use; such tools may, for example, help pupils draw pictures or graphs, or save information systematically. Programs of either of these types may provide a suitable context for teachers, or for pupils themselves, to pose problems for investigation.

The SMILE program, FACTORS (3), provides a game for a pupil or for a group of pupils to play against the computer. This game provides an interesting starting point for an investigation, which can be set by a teacher to pupils who have just played the game. No attempt will be made here to describe the game; it is difficult to describe briefly in words, but easy to grasp when presented in the form of a computer program.

Games for electronic calculators can similarly provide starting points for investigations. The SMILE booklet, *Calculating* (4), contains a problem called '1000 up'.

Can you make your calculator display the number 1000 by using only the keys below?

2   7   x   —   =

Many different questions can be asked following a successful solution of this problem. Here are some of them.

(a) What is the least number of key-presses needed to get 1000? How do you know?
(b) Does 999 take fewer or more key-presses?
(c) If you can choose which five keys you are allowed to press, what choice would you make in order to obtain the numbers 987, 654 and 321 with as few key-presses as possible?

The calculator does not play a major role in the solution of such subsidiary problems, but it plays an indispensable role in posing them.

Adventure games are computer games on a grand scale, and can provide the means for posing situations for investigation. *L — a Mathemagical Adventure* (5) contains numerous situations which must be successfully investigated if Runia is to be rescued. Here, as before, the computer offers little help in the investigation of the problems but provides an indispensable context for posing some of them and an interesting context for posing others.

BRANCH and SEEK are programs which allow users to structure information. The motivation for structuring it is a guessing game which the computer is taught to play. Here the program motivates pupils to find appropriate or interesting ways of structuring mathematical (or other) information. *Micromath* 1, 2, (6) contains an article describing how BRANCH was used in this way with a class of infants.

TURTLES is a program that permits two or more turtles to be moved simultaneously round the computer's screen. The use of TURTLES in a middle school to instigate class investigations into transformation geometry is described in an article in *Micromath* 1, 2, (7).

CIRCLE PATTERNS (8) is a program that enables its users to draw on the computer's screen the patterns obtained when equally spaced, numbered, dots on the circumference of a circle are joined according to a particular rule. The diagram shows the result of joining twelve dots according to the rule $n \rightarrow n + 3$.

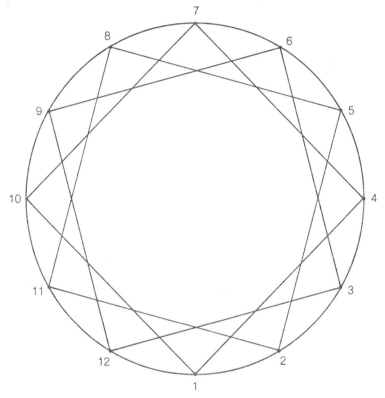

Figure 20. Diagram illustrating 'CIRCLE PATTERNS'

The program provides the context in which teachers can pose a number of problems for investigation. Here are a few of them.

(a) The example given above produces a pattern consisting of three squares. In what other ways can squares be obtained? In what other ways can three squares be obtained?

(b) When the rule $n \rightarrow 3n$ is used, patterns produced have one line of symmetry. Which numbers of dots must be used if the patterns are to have more symmetry than this? why?

(c) Investigate geometrically the patterns obtained using the rule $n \rightarrow -2n$.

A useful comparison can be made between the SMILE program CIRCLE (9) and the program CIRCLE PATTERNS. The latter enables the user to do much more than the former. The former can, however, be used by teachers without preparation, because it poses the problem for pupils to investigate. CIRCLE is therefore a program of the type described in the previous section, and has all the strengths and weaknesses of programs of this type.

Programs that support investigations

The previous section described how problems could be posed by the teacher, using the program CIRCLE PATTERNS. The program can also be used to support pupils when they begin investigating the problems. Pupils can save a considerable amount of time and labour by getting the computer to draw the diagrams required to test their hypotheses. The only diagrams they need draw for themselves are the key diagrams which are required to explain clearly the conclusions they arrive at.

   Other programs may not be indispensable for explaining the situation to be investigated (although teachers may sometimes find that they provide a handy 'electronic blackboard'), but they may, like CIRCLE PATTERNS, make it considerably quicker and less tedious for pupils to investigate that situation. TILEKIT (9) helps pupils decide which regular shapes, and which combinations of regular shapes, can be used to cover a plane surface. It also helps them explore some of the patterns which can be made by arranging shapes systematically, or by overlapping them.

   FUNCTION GRAPH PLOTTER (10) and other graph-plotting programs enable pupils to explore, with or without the help of teachers, the Cartesian graphs of polynomial, trigonometric and other functions. Here are some of the many problems that can be investigated.

(a) What does the graph of $y = 2x + 3$ look like? What is the effect of changing the 2? What is the effect of changing the 3?
(b) What does the graph of $y = 12 - 4x - x^2$ look like? What is the effect of changing the 12? What is the effect of changing the 4?
(c) What is the relationship between the graphs of $y = \sin x$ and $y = \sin 2x$? Explore the effect of other ways of complicating $y = \sin x$.
(d) Which polynomials' graphs are symmetrical?

None of these questions are particularly difficult to pose without a microcomputer. Some of them (and particularly the last) would be tedious to investigate in depth if all the graphs had to be drawn by hand.

A small minority of teachers has for some time used films to encourage pupils to investigate geometrical situations. Rather more teachers seem happier about using 'films' in the form of microcomputer programs, probably because most programs of this type are interactive and allow teachers and groups of pupils to ask questions which begin 'what would happen if?' ARMS and CIRCLES (10) are two programs of this type which allow users to explore properties of loci. One of the problems with film programs on eight-bit computers, such as the BBC microcomputer, is that the animation is usually slow and therefore the pictures have to be rather crude. The advent of sixteen-bit machines will make such programs far easier to write and may make it possible for geometrical investigations to play a greater part in the school mathematics curriculum.

Writing your own program to help you investigate a problem
In some mathematics classrooms teachers encourage pupils to write their own programs to help them learn mathematics. Such programs may be written in LOGO, in BASIC, or in another computer language. Frequently the purpose of these programs is to help pupils to investigate a mathematical topic.

Pupils in some primary schools make extensive use of LOGO, and of floor turtles in particular. Many of the tasks they set themselves, such as getting the turtle to draw pictures of particular objects of interest to them, are in effect mathematical investigations. Pupils can use LOGO to investigate more obviously geometrical problems. An article in *Micromath* , 1 (13) described in detail how two girls carried out such a self-determined investigation of this kind. The inexpensive booklet, *Creative Geometry* (14), provides several excellent ideas for geometrical investigations which involve pupils writing LOGO programs. Here is one of their problems.

Can you make a five-pointed star? How short can you make the procedure? Try making stars with other numbers of points.

Pupils also write BASIC programs to help them investigate mathematical problems. Sometimes such programs are suggested by teachers; at other times a pupil who has his or her own computer at home uses it naturally to help with homework. Among the problems which pupils have used computers to investigate in this way are Pythagorean triples, and the numbers that have exactly 15 factors. But a BASIC program is likely to be of help in any situation in which something has to be counted or a table of results has to be produced. Pupils may, for example, use a program to draw up the table of values required to draw a graph.

Other languages may also be useful. *Micromath* 1, 2, contains an article describing how PROLOG can be used to help solve a variant of the 'Frogs' problem, but PROLOG is a language which has so far been little used in mathematics classrooms, partly because few teachers have had access to the language. Spreadsheet programs may also have uses in mathematics classrooms in connection with investigations, and teachers are beginning to experiment with their use.

Programming as mathematical investigation

Before pupils can write a BASIC program to draw a house on the screen they need a working knowledge of coordinates; pupils who write a program to produce a number sequence or a table of numbers will almost inevitably learn some algebra or will come to understand better the algebra they know already. Pupils who want to write a BASIC program to draw a circle must first learn some elementary trigonometry. Pupils who write a LOGO program are incidentally building up their knowledge of angles, and of the angle properties of polygons, and may also be becoming aware, at an intuitive level, of important ideas in connection with limits or curvature. Thus almost any programming task undertaken by pupils can be thought of as a mathematical investigation.

Above all, pupils writing programs are learning from their mistakes. They are trying out ideas and finding that they do not work. They are then making hypotheses: what would happen if we changed that bit? Creating your own mathematics, learning to analyse what you have already done, learning from your mistakes and hypothesising are all activities which pupils engage in when they investigate mathematics. Pupils also inevitably become involved in all these activities when they write a program to make the computer do something.

Collaborative investigations

Teachers who watch pupils using computers in classrooms are constantly amazed by the length of time for which a group of pupils is prepared to concentrate on the task in hand and by the amount of purposeful conversation which takes place. Thus, one use to which computers can be put in mathematics classrooms is the support of group investigations. The advantages of encouraging a group of pupils, rather than an individual pupil, to engage with a mathematical problem are many. Pupils learn to hypothesise by listening to one another's hypotheses; they learn to reason by arguing against these hypotheses; they become more confident as they watch themselves and others publicly making mistakes which do not really matter because they can be corrected 'at the press of a button'.

With the help of computers, a teacher may find it easier to allow a class to explore a situation to which a pupil has drawn attention in the middle of expository teaching. For example, a teacher who is using a graph-plotting program to explain some aspects of graphs of a particular kind of function is more likely to respond favourably to 'what would happen if' questions asked by pupils, if the questions can be answered by a graph drawn by the computer on the screen rather than a graph drawn by the teacher on the blackboard.

With the help of computers, groups of children may also engage in group projects, or in individual projects where much of the initial work is done collaboratively. There is no reason why any of the programs described in this article should not be used by a group of pupils; indeed, an educational opportunity is likely to be lost whenever pupils are told to use such programs on their own.

Conclusion

Many mathematical investigations do not require a computer. Some investigations can make profitable use of a computer. For some investigations a computer is more or less indispensable. Computers may help some teachers more than others to present topics in an 'open' way and may provide for some pupils more than others the confidence to respond to such 'open' challenges. The question 'what will happen if?' is one of the commonest questions asked by the mathematician; it is also one of the commonest responses to an interactive computer program. It would therefore be surprising if computers did not make a major contribution to the quality of mathematics investigated in classrooms.

References

1. *Mathematics Counts,* Her Majesty's Stationery Office, 1982
2. This issue of *Micromath* contains three articles about this problem, and also a review of available software. *Micromath* is a journal of the Association of Teachers of Mathematics, and is published by Basil Blackwell
3. FACTORS is one of the programs in *MICROSMILE 1*, available from the Centre for Learning Resources, 275 Kensington Lane, London SE11 5QZ
4. *Calculating* is a booklet of activities for pupils using calculators, and is also available from the Centre for Learning Resources

5. Published by the Association of Teachers of Mathematics, Kings Chambers, Queen Street, Derby DE1 3DA

6. This is one of the software reviews

7. *A School of Turtles* by Barrie Galpin, *Micromath,* 1, 2, 1985

8. An unpublished program, available from the author

9. Another of the programs in *MICROSMILE 1*

10. One of the programs in *Some More Lessons in Mathematics with a Microcomputer,* published by the Association of Teachers of Mathematics (address above)

11. FUNCTION GRAPH PLOTTER is one of the programs in *Some Lessons in Mathematics with a Microcomputer,* published by ATM. It is also one of the programs in *Micros in the Mathematics Classroom,* published by Longman

12. These are both available in *Some Lessons in Mathematics with a Microcomputer,* published by the Association of Teachers of Mathematics (address above)

13. *Two Children and Logo* by Celia Hoyles, Rosamund Sutherland and Joan Evans, *Micromath,* 1, 1, 1985

14. *Creative Geometry: investigations on the microcomputer* is by Alan Bell, David Rooke and Alan Wigley. It is published by the Shell Centre for Mathematical Education, University of Nottingham

Developing mathematics from a micro

Ron Taylor, Boundstone Community College, Lancing

Since February 1984 I have been a member of a working party, funded by the Southern Regional Examination Board, whose brief was to look into the development of ideas and materials on a project approach to mathematics coursework and its assessment. The material being published is entitled 'Teachers evaluating and assessing mathematics'.

My role, in conjunction with the department, has been to produce a set of guidelines with examples of pupils' work, and classroom practice for initiating, sustaining and assessing investigations/projects. In September 1984 we introduced a new mode 3 CSE course as a pilot, through the SREB. This is based around mathematical activity in the classroom, and includes the assessment of three or four pieces of sustained work.

In addition the department has been looking into the use of microcomputers in mathematics. This second area of interest is linked with the MEP Curriculum Development Project in Mathematics, based at the West Sussex Institute of Higher Education. It is concerned with the application of 'small powerful' programs on the microcomputer for developing mathematical ideas and problem-solving ability. Initially the work was done with A-level pupils outside their normal timetabled mathematics lessons, but has now been extended to the whole ability- and age-ranges with extensive use of the computer within lesson time.

I have included extracts from two projects to illustrate the different ways of using a micro to encourage mathematical thinking. 'Clowning Around', a group project by four 13-year-old girls, and '3U + 2T' by a 13-year-old girl. All were from the same mixed-ability class.

'Clowning Around' developed from a small program for the BBC micro, which allowed the user to draw circles of any size and at any position on the monitor screen. The program was taken from an Association of Teachers of Mathematics activity book containing short computer programs. After they had entered it, they set themselves the task of drawing a simple face.

'3U + 2T' was started entirely through the pupil's own initiative; she had become fascinated by the 'number chains' obtained for different rules. She decided to investigate this particular rule (multiply the units digit by three and add twice the tens digit), as it was to her 'the most interesting'. The micro was used to alleviate the tedium of a large amount of arithmetic and to facilitate further research.

'Clowning Around'

This is the initial program that was presented to the pupils.

```
20 MODE 4
30 VDU 28,0,3,5,0
50 INPUT X,Y,A
60 FOR K = 1 TO 120
70 LET L = K/60 * PI
80 PLOT 69,X − A * COS(L),Y + A * SIN(L)
90 NEXT K
```

They spent about two weeks working on the computer before they were satisfied with their changes to the initial program. There were several pages of rough working and lots of trial and error.

This group of pupils was well organized and divided the task among themselves very well: two on the planning and drawing; two on the input into the computer. General discussion took place when further changes were sought, often stimulated by the nature of the image on the screen. The work done by pupils is activity-based with a flexible routine. The lessons are organized in such a way as to give pupils the opportunity to continue or return to a piece of work they find interesting. The climate in the classroom encourages pupils to tackle problems in their own way, the mathematics arising naturally out of the activity. The pupils enjoyed the project (spending six weeks on it) and still talk of adding further extensions such as filling in the clown's hat.

The pupils were not taught any BASIC. The need to draw a 'Clown' generated the need to research and learn more about the workings of the computer which is itself an investigative and productive activity. This was not necessarily revealed in their write-up of their project nor in the extracts that follow.

Figure 21 shows the first page of the project. When establishing the size of the screen and what the variables represented in line 50, there was a great deal of trial and error. If a mistake was made the whole screen had to be wiped clean. The initial reason for adding line 55 was to obliterate a single circle by going over it in black. Later this particular instruction was used to produce a more colourful face, namely that of a clown. Variables B and S replaced numbers 1 and 120 in line 60 to draw part of a circle. In the first instance this was done so that they could draw a mouth.

Figure 22 is their own construction of the final clown's face, which they drew on graph paper. Figure 23 is a screen dump. Certain colours such as green and blue did not transfer and these have been drawn by the pupils.

Clowning Around.

We were given a program that drew circles anywhere on the screen. We adapted it so that we could draw circles, any amount, any size and any colour. We are drawing a clowns' head with this program.

This is the program:

```
20 MODE 2
30 VDU 28,0,3,5,0
50 INPUT X,Y,A,B,S,C
55 GCOL 0,C
60 FOR K=B TO S
70 LET L=K/60*PI
80 PLOT 69 ,X-A*COS(L),Y+A*SIN(L)
90 NEXT K
100 GOTO 50
```

We used SIN and COS in this program. SIN means UP and COS means ALONG.

When we have typed the program in, a question mark comes up on the screen 6 times. We have to type in:

1. How far along we want the circle.
2. How high up.
3. How big we want the circle
4. Where the circle starts.....
5. and ends.
6. What colour we want it to be.

When we make mistakes and we put circles in the wrong places. We don't need to start again, we just go over the circle in black and it doesn't show because the screen is black.

Figure 21. First page of the 'Clowning Around' project

The final listing is on p98. Retyping in data was time-consuming, so I introduced the pupils to the READ and DATA statements. Initially they added a GOTO 50 in line 100, crude but effective. In order to draw a hat using MOVE and DRAW after the face was drawn required the introduction of the FOR NEXT loop in line 40. The brim of the hat is drawn in line 360.

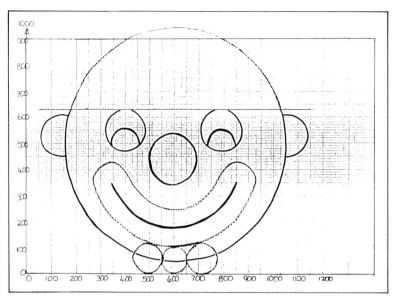

Figure 22. Final construction of clown's face drawn on graph paper

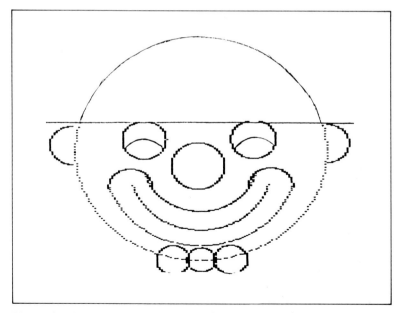

Figure 23. Screen dump of clown's face

```
 20 MODE 2
 30 VDU 28,0,3,5,0
 40 FOR D = 1 TO 22
 50 READ X,Y,A,B,S,C
 55 GCOL 0,C
 60 FOR K = B TO S STEP .5
 70 LET L = K/60 * PI
 80 PLOT 69,X − A * COS(L),Y + A * SIN (L)
 90 NEXT K
100 NEXT D
110 DATA 600,500,450,0,120,7 :REM FACE
115 DATA 850,350,80,0,60,1
120 DATA 350,350,80,0,30,1
130 DATA 850,350,80,0,60,1
140 DATA 350,350,80,30,0,1
180 DATA 140,530,80,90,120,7
190 DATA 140,530,80,0,30,7
200 DATA 1050,530,80,30,90,7
210 DATA 592,440,95,0,120,1
220 DATA 600,450,200,65,115,1
250 DATA 350,350,80,30,60,1
260 DATA 400,550,80,0,120,7
270 DATA 790,550,80,0,120,7
280 DATA 600,450,340,65,115,1
290 DATA 600,425,250,65,115,7
300 DATA 400,500,60,0,60,4
310 DATA 790,500,60,0,60,4
320 DATA 600,1000,450,60,0,2
330 DATA 600,50,50,0,120,3
340 DATA 500,50,60,0,120,9
350 DATA 705,50,60,0,120,9
355 DATA 600,500,450,5,55,2
360 MOVE 40,630:GCOL 0,2:DRAW 1160,630
```

The following pieces of work were written by different pupils. It is their attempt to explain why each stage of the program developed and how it worked.

Mary
'We used this program to draw most of the clown's face.
The eyes, ears and the bow tie are made of whole circles but often half-circles, etc, need to be drawn.

A circle on the computer is divided into four main points.

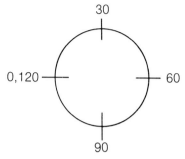

So if for example you need a half-circle, the fourth and fifth numbers that you would type out would be 0,60. To draw three-quarters of a circle you would type out 0,90, etc.

After we had worked out most of the clown's face we changed the program. These were the lines we added.

```
40 FOR D = 1 TO 22
50 READ X,Y,A,B,S,C
100 NEXT D
```

Then we typed out each set of numbers for the clown's face. The first numbers were as follows:

```
110 DATA 600,500,450,0,120,7.'
```

Jane
The following is Jane's explanation of how they drew the clown's hat. 'One of the last things we added to it was the straight line representing the brim of the hat. To be able to draw a straight line in this program we had to stop the data lines and add another statement. To do this we had to change line 40 from:

```
40 READ X,Y,B,S,C to
40 FOR D = 1 TO 22
```

The change allows the computer to read all 22 data statements and then obey any other following statement.

So we added a simple MOVE and DRAW statement.

To get the top of the hat in the right place (exactly on top of the brim both ends), we had to calculate how far around the circle each end of the edge of the hat was.

The position of each end of the hat was roughly about 6 squares up from 0 and 6 squares up from 60.'

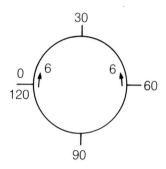

Susan

These are some of Susan's descriptions of the data lines. 'These are the numbers we drew to make the clown's face.
The head: Data line 110

600 – How far along the screen the centre of the circle is
500 – How far up the screen the centre of the circle is
450 – How big the circle is, each direction from the centre
0 – Where the circle starts
120 – Where the circle finishes
7 – What colour it is

The left ear (two parts) lines 180 and 190

140	140
530	530
80	80
90	0
120	30
7	7

'Clowning Around': some observations
In order to determine the nature of the mathematical learning that took place while these pupils were involved in this project, let us try to list those strategies/processes employed and the mathematical content generated.

The following is a list of some processes and strategies that were used by these pupils while engaged in this project.

(a) Keeping one variable fixed while changing others
For example when determining what the variables X, Y, A stood for in line 40.

(b) Deciding what to exclude, what to consider
When first investigating this program, the pupils had to make decisions about which part of the BASIC coding they would attempt to understand. They established that lines 50 to 80 drew a circle, but left out any detailed analysis.

(c) Spotting blind alleys and dead ends
The pupils made many conjectures in trying to understand the function of each line of the program, and decisions had to be made as to when to cease a certain line of inquiry.

(d) Being systematic and organized
The need for this was generated within the group.

(e) Using guesswork and intuition
A very strong force initially and an underrated process.

(f) Prediction
Once they understood how the program worked and the size of the graphics grid, they were able to predict and determine the exact size and location of each circle.

(g) Defining further problems, or extensions
In order to create a more realistic image, and remove the burden of typing in the data on each separate occasion, extra statements were added:

(i) GCOL 0,C was to allow for the removal of previous trials
(ii) validate B, to S, to draw a more realistic shape, eg mouth, ears
(iii) use of the READ and FOR NEXT loop to save re-entry of data.

(h) Logical reasoning
Deducing the action of SIN and COS without any 'teaching' on my part.

Since it was a group project there were several other important aspects. I have included references to those sections from the 'Aims' in the 'National Criteria for Mathematics' which I think are covered by this project:

Co-operation between pupils
Splitting up of tasks
Ability to communicate clearly both orally and in writing, cf 2.1, 2.2
A planned line of action
Research and reporting back: for example, when determining the size of the clown, two of the pupils worked on graph paper before entering their results into the computer, cf 2.3, 2.5, 2.6, 2.7, 2.8, 2.13

An ability to interpret a 'foreign code'
In addition it can be readily seen that the following 'Assessment objectives', outlined in the 'National Criteria for Mathematics', are also apparent in 'Clowning Around', viz, 3.2, 3.3, 3.4, 3.7 to 3.14, and 3.16, 3.17.

The following is a list of some of the mathematical content:
(a) interpreting the BASIC computer language
(b) using drawing instruments accurately
(c) understanding and applying the two-dimensional coordinate system
(d) scaling
(e) symbolism: understanding and developing their own
(f) estimation and approximation; working to an appropriate degree of accuracy
(g) performing calculations, using fractions
(h) using terminology relating to circle, eg, arc, circumference, etc.

Which of these two lists of outcomes do I value more highly? Certainly, to date I feel there has been far too much emphasis placed upon the content list in a syllabus, and too little attention paid to its aims. Mathematics should encourage and help children develop those strategies and processes outlined above, as well as offer them a great deal of enjoyment. When this happens the content emerges as a by-product, and those anxieties and stresses so often associated with failure, when undue emphasis is placed on skill and technique, are no longer present.

3U + 2T
Figure 24 shows part of the first page from Samantha's project.

This investigation began as an ordinary problem trying to find out whether a pattern could be seen when the Units number was multiplied by 9 and the tens number was added. This was interesting in that nearly every number was used when the starting number was 2. But I decided to investigate other such equations. 3U+2T was the most interesting.

3U+2T is an equation of the units number multiplied by three added to twice the tens number (eg)

Units number would be the 8 of 18 — 18
Tens number would be the 1 of 18 — 18

So if 18 was a starting number 3U+2T would follow:-

Starting at 18
8×3+1×2= 26

Figure 24. First page of Samantha's project

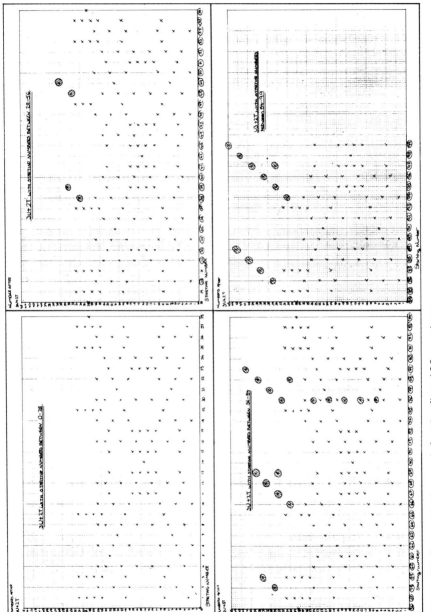

Figure 25. Samantha's graph for the first 100 numbers

The first 28 numbers were done on paper and the results plotted on a graph (this was her own decision). She was intrigued by the shape of the graph (Figure 25) and wanted to do all the numbers up to 100. This would be time-consuming, and using the computer would remove the tedium of a large amount of arithmetic as well as facilitate further research.

Samantha had done some programming before and was familiar with the BASIC commands INPUT, PRINT, IF, THEN, and GOTO. I then introduced MOD and DIV. Below is the program which we wrote together. The overall structure of the program is hers.

```
10  INPUT NUMBER
15  FIRST = NUMBER
20  UNIT = NUMBER MOD 10:TEN = NUMBER DIV 10
30  NNEW = 3 * UNIT + 2 * TEN
40  PRINT NNEW
50  IF NNEW = FIRST THEN 10
60  NUMBER = NNEW
70  GOTO 20
```

These are some of her results. The notation is her own.

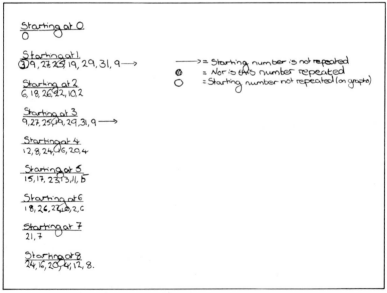

Figure 26. Some of Samantha's annotated results

The following are some of her many conclusions.

'All of the graphs are basically symmetrical.

Numbers 1 to 29 and 31 are repeated.

The only number which occurs after applying the rule "3U + 2T" to 14, 42, 70 and 98, is 14.

The only number which occurs after applying the rule to 28, 56, and 84, is 28.

38, 39, 48, 49, 58, 59, 66, 67, 68, 69, 76, 77, 78, 79, 86, 87, 88, 89, 94, 95, 96, 97, 98, and 99 are not repeated. This is because thrice the units digit plus twice the tens digit is never any of the numbers from 1 to 29 or 31. So the second number will not be repeated either.

Numbers that do not repeat have the same number chains as those numbers 28 less than them eg,
$55 - 25 - 19 - 29 - 31 - 9 - 27 - 25$ and compare this with the number chain for 27 $(55 - 28 = 27)$:
$27 - 25 - 19 - 29 - 31 - 9 - 27$. So 2, 30, 58, 86 have similar number chains.'

These conclusions are a rich source for further study. Explanations and generalizations are made easy with the use of algebra. Its use grows out of the investigation, and can be related at some later stage in her mathematical development. For example, take any two-digit number ab (a = tens digit, b = units digit), and ask 'which numbers cycle onto themselves?' For this to happen we want $3b + 2a = 10a + b$. That is $2b = 8a$, or $\frac{a}{b} = \frac{1}{4}$. Answer 14 and 28.

The need to symbolize, use shorthand and learn some new BASIC commands, arose naturally.

The graphing of her results, which show a high degree of symmetry, was entirely her own idea. Samantha had worked at this investigation for some time before she asked me to help her write a program to investigate higher numbers. The program aided further research, and acted as a check on a number of hypotheses she had made, and encouraged new conjectures.

A similar exercise to that carried out for 'Clowning Around' is easy to do, but I think the work speaks for itself.

References

TEAM: Teachers Evaluating and Assessing Mathematics Materials available from: Mathematics Education Centre, West Sussex Institute of Higher Education, Bognor Regis, West Sussex. (The materials contain the examples discussed in the article and are reproduced by kind permission of the West Sussex Institute of Higher Education.)

MEP: the Microelectronics Education Programme at the West Sussex Institute of Higher Education (Project Director: Adrian Oldknow)

ATM Activity Book: *Problem-Solving Workcards For Use With Microcomputers* by Jim Seth, Association of Teachers of Mathematics

Algorithms

Adrian Oldknow, Mathematics Curriculum Development Centre, West Sussex Institute of Higher Education

Introduction

The word 'algorithm' has become more and more widely used, probably as a result of the development of computing devices. Yet, in common with other bits of jargon such as 'interface' and 'parameter', I suspect that it does not convey much meaning to most people. We joked about the computer out-crowd who used to think that Research Machines 380Z microcomputer was the 38oz (thirty-eight ounce) micro, but I guess there are a fair number of people who see the word 'algorithm' as a misprint for 'logarithm'. Why, then, should people with an interest in teaching and learning mathematics need to bother their heads now about algorithms when they managed quite well without knowing about them in the past?

I first came consciously face-to-face with algorithms in the context of computer science when I started to read a book by Donald E Knuth called *The Art of Computer Programming, Vol. 1: Fundamental Algorithms'*, published by Addison-Wesley. I confess straight away that I did not learn very much about programming computers from this book, but it did open my eyes to a lot of stimulating mathematics and its history. For a start it told me what an algorithm was and it made me realize that the 'flowcharting' I had been teaching from SMP was just one way of representing an algorithm.

Later I came across the work of the German mathematical educator, Arthur Engel, including a book titled: *Elementarmathematik vom algorithmischen Standpunkt*, published by Ernst Klett Verlag. Fortunately an English translation of this has now been produced by F R Watson and is available from Keele Mathematical Education Publications under the title *Elementary Mathematics from an Algorithmic Standpoint*. Engel updates Felix Klein's approach to 'functional thinking' to the present era, with its access to calculators and computers, under the slogan 'algorithmic thinking'. He contends that the concept of an algorithm should be the key concept in school mathematics and that we need to reconsider the content of school mathematical syllabuses from an algorithmic standpoint.

In May 1983 the Department of Education and Science and the Microelectronics Education Programme jointly sponsored a Curriculum Conference at Pendley Manor, Hertfordshire, to advise on policy for future curriculum developments in a number of secondary school subjects involving microcomputers. The report of the mathematics group, under the chairmanship of Trevor Fletcher, was published in the March 1985 edition of *Mathematics in Schools*. This report places particular emphasis on 'mathematical programming', 'algorithms and procedures' and 'short programs' as shown in the following extracts.

'We consider that mathematical programming should be a staple part of mathematics courses in the future. Just as a calculator has to be on hand if arithmetic is to be taught to best advantage so a computer will be needed if algebra, geometry, statistics and other branches of mathematics are to be taught to best advantage.'

'In the light of advances in technology primary and secondary school mathematics should include the explicit study of algorithms as an integral part of the curriculum. This means that pupils will need to be involved in all stages of the study, ie, the performing of given procedures, the modification of procedures, the design and evaluation of new or alternative procedures. This requires the availability of suitable tools from pencil and paper and concrete apparatus to calculators and computers, and a full appreciation requires that pupils make some use of a suitable programming language . . . eg, BASIC and LOGO.'

'The study of algorithms and procedures represents both an extension of mathematics and a new way to view the current school curriculum. In addition an algorithmic approach provides a powerful facilitator in the development of children's understanding of mathematics. Mathematics teachers have traditionally taught children to apply specific well-known algorithms (for example: long multiplication) often by rote, but the children are seldom, if ever, given the opportunity to explore, modify or design algorithms — an important feature of mathematics in a technological society.'

'At the other extreme is the 5 or 10 line program, developed in class, input by the teacher with pupil cooperation, or by the pupils themselves. Within the constraints imposed by whatever programming language is used, the user has complete freedom. Programs will generally not produce appropriate results first time, indeed the process of correcting errors of logic or of modelling (rather than syntax) is a central part of the learning of the mathematical concepts under study.'

More recently in a series of discussion documents on the school curriculum produced by Her Majesty's Inspectors of Schools *Mathematics from 5 to 16* stresses the need for pupils to have the necessary access and skills to be able to use microcomputers in mathematical investigations, using both appropriate software and programs of their own. It goes on to say:

'Microcomputers are a powerful means of doing mathematics extremely quickly and sometimes in a visually dramatic way. If pupils are to use microcomputers in this way they will need to learn to program the machines and so, if programming is not taught elsewhere, it should be included in mathematics lessons. For mathematical purposes such programming does not need to be highly sophisticated. It may be a form of LOGO, or the early stages of a language such as BASIC, or indeed any form of computer control which enables pupils to carry out their own mathematical activities.'

So there is some documentary evidence to show growing support for the *policy* of the inclusion of an algorithmic approach (and elementary mathematical programming) in the mathematics curriculum. But, as ever, there remains the not insignificant task of how best to *implement* such a policy. Of course the first constraint on any such curriculum development is hardware. Schools are short of appropriate resources in general and microcomputers are not readily available for the majority of secondary-school mathematics departments.

The Pendley Manor report was also keen to see the development of a small, low-cost, portable computer for use by individuals which had a good graphics capability and was easy to use. Such machines are now beginning to come on the market: there is now a range of personal microcomputers, using BASIC, about the size of an ordinary calculator and costing about £50 but, as yet, without good graphic displays and there is a brand new programmable calculator with a graphic display (about 100×100 pixels) at about £85. Thus we are already very close to having the kind of micro available that the Pendley paper envisaged and, given recent trends in the pricing of electronic devices, it is quite feasible to envisage a mathematics department having 15 or 20 such micros for less than the current price of a single BBC system with disc and monitor (and without the problems of trailing wires, flickering screens, etc).

If we can accept that hardware need not be a major constraint then we are left with the other issues such as the availability of suitable curriculum materials, changes in the examination process and in-service training for teachers. The rest of this article, then, is addressed to the first of these issues and describes some possible approaches to introducing 'algorithmic thinking' in mathematics lessons.

The approach envisaged is *not* to have a stock set of algorithms that are 'gone through', but to raise situations where algorithms are needed and in which pupils or students propose ones of their own. From this can come discussions about which are easier, more efficient, more robust, more general, etc. A major objective is for all of us to improve our ability to verbalize the techniques and strategies that we use ourselves, to make us more conscious of the strategies and algorithms we tend to use, and to question why we have come to use them. Remember the warning in Pendley that this is not concerned with the learning by rote of particular algorithms — the distinction between these two approaches to algorithms is highlighted in Stephen B Maurer's article: 'Two meanings of algorithmic mathematics' in the September 1984 edition of the *Mathematics Teacher*, the journal of the American Council of Teachers of Mathematics.

1. 'Binary Chop'

This is a powerful algorithm that can crop up easily in a variety of situations. The 'Number Guesser' game on the Texas Instruments' hand-held 'Dataman' is an implementation that is directly appropriate for the algorithm and computer programs such as 'Hunt the Hurkle' or 'Pirates', which are two-dimensional coordinate guessing games, invite a similar strategy. I use the 'Dictionary Game' as my starting point (this is described more fully in *Learning Mathematics with Micros* by Oldknow and Smith, published by Ellis Horwood). It starts by getting a class to try to guess my word in 20 questions — they always fail! I then do the party-trick of finding their chosen word (so long as it is in the dictionary) in about 15 questions, all of the form: 'Does your word come before the word "sausage" in the dictionary?' (*Very* boring.) In this way we refine down successively smaller intervals, each of which contains the unknown word.

The remarkable property is the relationship between the number of questions, Q, and the number of items, N, that can be distinguished in this way: $N = 2^Q$. Thus 10 questions allow $2^{10} = 1024$ (or 1K or about 10^3) items to be identified. With 20 questions you can locate one item in a *million*. Although this is an example of a well-known phenomenon, exponential growth, it is the basis of many 'public bar' conundrums, like the number of grains of rice on a chessboard: 'I put 1 on the first square, 2 on the next, 4 on the next . . .

all the shipping in the world couldn't carry the rice'; ($2^{64} - 1$ grains are about 20 million million million or $2 * 10^{19}$). For the 'Number Guesser' game suppose that we are looking for a number between 1 and 100. Now 100 lies between $64 = 2^6$ and $128 = 2^7$ so we ought always to be able to get it by using a binary chop of seven questions or fewer, see program below. One of the nice things about this algorithm is that once seen it is hard to forget and you do not really need to have to worry about finding words to 'say it to yourself'.

There is a nice cross-curricular tie-up with taxonomies in biology and botany — where the conventional notation system uses numbered questions and an implied IF ? THEN GOTO X ELSE GOTO Y structure — a true spaghetti plant. Suppose you have 10 things to distinguish (man, parrot, elephant, spider, crab, fish, butterfly, crocodile, oyster, frog) with a set of questions. Using an algorithmic approach we can make the strategy explicit — 'What sort of initial split do you want: 5/5, 4/6, 3/7, 2/8, 1/9? What are the pros and cons of each? Can you think of a question that splits in your chosen way? Start to make out a 'tree-diagram' showing the way the initial set is divided into subsets. What is the least 'depth' such a tree may have?' The *science* is in identifying what are sensible and useful types of questions — in order to distinguish between an elephant and a fish the question: 'Is it usually served with chips?' is not of great scientific significance. Perhaps this kind of link is easier to make in the primary curriculum (using, for example, the 'Animal' program in the MEP Microprimer pack).

```
10   Number = RND(100)
20   REPEAT
30      INPUT Guess
40      IF Guess < Number THEN PRINT "Too Small"
50      IF Guess > Number THEN PRINT "Too Big"
60   UNTIL Guess = Number
70   PRINT "Well Done"
```

2. 'Russian Peasant Multiplication'
I 'teach' this by saying: 'I have got a set of rules that I am going to follow, and I am going to do some worked examples on the board without speaking. You have to try to fathom out what I am up to. If you think you have spotted what I am doing then can you suggest an example to check that my way and your way agree.' In other words I try to emphasize the process of *inducing* a set of rules as the common structure on a set of disparate examples.

My teaching skill is in choosing the examples well and writing clearly (and not making any mistakes too early). The process has to be seen as *dynamic*. It could be presented on video but it is not much use my just writing down the end-product and giving it to you as a static image on paper. Some of the rules are usually spotted quickly — 'you're halving one and doubling the other', 'you stop at 1', but most people do not see what the algorithm does, how the numbers to be added are chosen and especially *why* it works. Just a little 'mutter' can soon let the cat out of the bag with words like 'multiply' and 'odd', but '*why*' usually defeats even able pupils.

I then try to get people to write down, or say out loud, what the steps of the algorithm are. Here we might introduce flowcharts and associated algebra (as in SMP) or some simple programming language or any other convenient algorithmic representation (such as the structure diagrams used in conjunction with PASCAL). Currently I am happy to use bowdlerized BBC BASIC, eg: 'To multiply two numbers, REPEATedly halve the smaller, double the larger, recording both UNTIL the smaller is equal to 1, then add up the larger numbers corresponding to smaller numbers which are odd'.

The 'why' is just too simple when the idea of 'equivalent products' has been pointed out, and no formal symbolic proof is needed. Then I like to say: 'I have just taught you to multiply — why weren't you taught this way in the first place?' We play at being the Board of Directors of an educational supply company and try to think up slogans for why customers should abandon the current market leader. 'Long Multiplication', and buy 'our' product instead. It does not take long before the tie-up with 'Learning Tables' takes place. I then tell them (probably a lie) that it was taught in Russian elementary schools at the time of the Czars as being good enough for most to cope with the kind of multiplication that peasants need. If some smart Alec (or Alicia) pipes up: 'But why do we need to worry about any multiplication algorithm now that we have got calculators' then we really have got something worth talking about.

The hardware arithmetic operations in calculators and computers are performed in binary, and Russian peasant multiplication is equivalent to binary long multiplication using 'shift' for doubling. Thus without 'Russian peasant multiplication' you might not have your calculator! We can then go through one of the examples again using the equivalent binary 'long multiplication'. It is interesting then to find out what algorithms different people use for multiplying in different situations — how do *you* multiply by 98?

Given the 'REPEATedly . . . UNTIL' form of expression the structure of the algorithm is explicit. To write a computer program we have just to decide how to encode words like 'halve', 'double', 'odd' and to think about sequential or parallel processes. By the way, the definition of 'oddness' or 'evenness' is in itself interesting. 'Evenness' means 'leaves no remainder on division by two', and 'oddness' means 'not even'. However we know that a number is divisible by two if its last digit is divisible by two, and thus we have an alternative test: 'a number is even if its last digit is 0,2,4,6 or 8'. This would not be a good algorithm for use in a simple computer program — why?

```
10   INPUT First,Second
20   Total = 0
30   REPEAT
40      First = First DIV 2
50      Second = 2 * Second
60      IF (First MOD 2) = 1 THEN Total = Total + Second
70   UNTIL First = 1
80   PRINT Total
```

3. Decimal to binary conversion

I use one algorithm, yet teach another — why? My own algorithm uses 'top down' decomposition into sums of powers of two: 'REPEATedly find the largest power of 2 smaller than or equal to the number and replace the number by the difference UNTIL the number is 1 or 0' but I guess it is easier to teach a 'bottom-up' method: 'REPEATedly replace the number by half of itself, noting down the remainders UNTIL the number is 0'.

```
10   INPUT Number
20   REPEAT
30      Digit = Number MOD 2
40      Number = Number DIV 2
50      PRINT Digit
60   UNTIL Number = 0
```

There is a temptation here to want to use strings; how about binary to decimal?

4. Euclid's algorithm

This is supposed to be the 'classic' example of an algorithm. But once again it is not often met in mathematics classes. Again I like to use the approach of doing sums on the board without talking to see if the class can guess my process. Here there is plenty of scope to talk about refining the algorithm to make it more efficient, varying the UNTIL condition, using division instead of repeated subtraction and so on.

REPEATedly replace the larger by the difference UNTIL both are equal.

REPEATedly replace the larger by the difference UNTIL either both are equal OR the smaller is 1.

REPEATedly replace the larger by the remainder of the larger divided by the smaller UNTIL remainder is 0.

The 'why' is also a good question. If a number is a factor of two other numbers then it is also a factor of their difference: enough said? Is that a proof? How were *you* taught to find the highest common factor (HCF) of two numbers? I learnt to put each into prime factored form (an algorithm in itself) and then to pick common factors — yet I am quite sure that is not how I do it. The nice thing about this is the tie-up with the least common multiple (LCM): LCM(a,b) = a $\star$ b/HCF(a,b) — which is a frequently used computer algorithm.

I suspect that one criterion used in the selection of algorithms suitable for Victorian elementary education was that all the working should fit on to one side of a B5 slate!

Here, then, are just a few simple examples from my own attempts to implement an 'algorithmic approach' in mathematics. I have concentrated on algorithms, rather than programming, and I would thoroughly recommend reading the booklet *Programming in Primary Schools?* which is part of the MEP Primary In-Service Pack on *Problem Posing and Problem-Solving.*

Teaching and learning mathematics — will programming help?

David Johnson, Shell Professor of Mathematics Education, Centre for Science and Mathematics Education, Chelsea College (King's College), University of London

This article first appeared in *Programming in Primary Schools? A Course Reader* which can be found in the MEP Primary In-Service Pack on *Problem Posing and Problem Solving*

Introduction

To begin let me say that the title of this paper is really only a rhetorical question as we already have considerable evidence which tells us the answer is yes. One other point to be made is that the inclusion of the word 'mathematics' in the title is crucial. This is so that the arguments which follow can be placed in a school context. I am not proposing to get into the debate as to the merits of one programming language over another (the usual squabble) as such discussions or arguments are usually at the level of considerations of programming power and language features *per se,* often at a rather esoteric level, rather than in terms of *accessibility* and *adequacy* for the proposed task, in this case 'doing' some mathematics (and this is more than 'turtling'). With this in mind let me place my views before you in the form of some points for discussion (this is a modified version of the 'Notes for parent or teacher' in my book *Explore Maths with your Micro: a book for kids aged 9 to 90).*

Some points for discussion

A. Studying mathematics

1. Mathematical concepts and relationships become powerful tools for exploration and investigation when viewed as *dynamic* entities rather than static definitions or statements.

2. *Discovery* enhances learning.

3. *Interaction* and *discussion* are among the most, if not *the* most, important aspects of teaching and learning; and in a computing environment this means

(a) the individual child interacting with the machine; and

(b) even more important, children working with the machine and talking to one another about what has happened and what to do next.

4. Mathematics is a *formal system* — one which requires both precision and rigor. This is truly reflected in an approach which includes computer programming.

B. Computer programming in mathematics
1. There is considerable support, research findings and personal opinion, that programming is best learned by following a sequence of activities which involve:

(a) using 'blocks of code' or short programs to do a significant and/or interesting task: in such a situation the children will see the 'words' working
(b) modifying or extending a short program to do a new but related task
(c) and finally, designing or preparing new programs to solve problems.

2. The computer is only a tool or device for extending the intellectual power of the user.

3. The choice of computer language, for example BASIC, LOGO or PROLOG, is *not* the important issue; what is important is the identification of appropriate tasks or problems (explorations and investigations).

C. The 'classroom' — this may be home or school
1. Children should work together and share ideas.

2. Keeping a copy of each program, with notes and output, is crucial.

3. Extending the work — new ideas should be explored and all new ideas are good ideas.

I would be unhappy if you did not choose to debate some of these points, at least at the level of asking for further clarification. What are your views on A3? B3? C3? The present paper is far too short to consider all the points, hence I propose to concentrate on A1, A4, B1, and B3. In the case of B3 I will note here that while computing, or better yet 'information-processing', has implications and potential for use in all subjects, ie, computing is more than mathematics, it is also the case that mathematics has a unique link, both historically and in terms of the role of programming in the study of the subject. By mathematics I include both processes and content and suggest that with the computer the child can study and observe patterns and investigate proposed or hypothesized relationships. Hence, as stated in the opening paragraph I propose to restrict the discussion which follows to the subject domain of mathematics — with no apologies.

Mathematical ideas as procedures

A number of research studies over the past 20 years have shown that certain mathematical concepts or relationships are better learned when children have been asked to imbed the concept in a computer program. However, it is only recently that we have come to realize that it is not really a case of 'imbedding in a program' but rather the situation is that of the concept taking on a very different character altogether — *the concept can be viewed as a procedure*, ie, an ordered sequence of steps for doing a particular task, and hence a dynamic entity rather than a static definition or statement. (Note: I will use the words procedure, algorithm, and 'computer program' interchangeably throughout this paper although the latter is really just a means or form for communicating or describing an algorithm or procedure — two other forms are a list of directions — sentences — and a flow diagram. The computer program however is also a form which can be processed by machine.) To return to the notion that certain concepts or ideas become more useful, or 'powerful', when viewed as procedures, let's look at some examples.

Prime numbers

My favourite example for demonstrating the difference between a mathematical idea viewed dynamically, as a procedure, as contrasted with a static definition, is that of a prime number. You will remember that a prime number is a number whose only factors (divisors) are one and the number itself. Hence, 2, 3, 5, and 7 are all the prime numbers less than 10; and children who have learned about prime numbers can usually, with a little thought, give you this set. But, ask these same children 'Is 57 prime?' or 'Is 51 prime?' and a common response to both questions is 'yes'. Why is this? I suggest the reason is that while the children may know the definition (given above), they do not have an effective means of applying this definition and hence merely look at the number and try to recall a basic number bond. When they fail to think of a number pair they then come to the conclusion that the number is a prime number (and who can blame them, for doesn't 51, or 57, look prime? — each has a primeness 'feel' or appearance, at least they do to me).

If one reflects on this situation, it soon becomes clear that to determine whether or not a given number is prime requires one to think of 'prime or not prime' as a procedure. The children should be given the opportunity to actually design the procedure. They can then use a calculator to test their ideas with some numbers, say 7, 10, 51, and 901. Their procedure will quite likely be based on the ideas included in the procedure given below.

To test for primes (with a calculator). Divide the number by 2, then 3, then 4, and so on up to one less than the number. If any division gives a whole number as a result *stop* and write down '(the number) is not prime and has (the divisor) as a factor.' Otherwise write down '(the number) is prime'.

This activity is not intended to replace the early look at the famous 'Sieve of Eratosthenes' (Eratosthenes, 276 – 196BC) or any of the other interesting activities now used to introduce the idea of a prime number. However, it is the case that expressing the concept as a procedure enables the child to use it to investigate other classical relationships such as 'the frequency of prime numbers in consecutive decades' (or centuries) or the quite fascinating theorem that 'between any positive whole number and its double there is at least one prime number' (proposed and proved by Chebichev (1821-1894), one of Russia's most distinguished mathematicians).

Enough about prime numbers, for now. This example was selected to be presented first because the distinction between the idea of a static definition and a dynamic procedure is quite easy to observe. On the other hand you might be asking whether such a distinction holds for more elementary ideas and further, where the programming is in this (although you should now have some ideas about developing a program to test for primes even if you are not sure about the coding). Let's go back and consider some of the mathematics which precedes any consideration of prime numbers — how about multiples? and factors?

Multiples
A child's view of a set of multiples will quite likely be restricted to a consideration of a specific set and the representation of the set (and the 'counting pattern'). That is, a set of multiples is thought of as 'the multiples of 3': the set 3, 6, 9, 12, etc; or 'the multiples of 5': the set 5, 10, 15, 20 etc. Thus one might conclude that this is very much an example of a static description, and each set 'stands on its own'. The particular set is the item of concern. Now think about what it means to approach this idea as a procedure, that is, the concept of

a set of multiples as a way of getting the multiples and ultimately getting any number of multiples for any starting number. An approach I have used, after the children have spent considerable time working on the ideas and models related to multiplication (multiplication as modelled by successive addition, the array, and Cartesian product) and ample practice with single-digit number bonds, is to present the children with a short BASIC program to generate multiples of 5, such as that given below (note: if the children have done some programming in BASIC or LOGO we might develop this as a small-group or class exercise or they might be left to work in groups of two or three and write their own program).

BASIC and LOGO versions are included for the reader to illustrate that either is suitable. The actual choice can depend on such things as what machines are available (for example the Sinclair ZX81 is quite suitable for work in mathematics, but only has BASIC) and/or what previous exposure children have had to programming — if they have not done much with LOGO then BASIC is probably easier to use for number work such as that included in this paper. One final point: it is also the case that there is nothing wrong in seeing and using more than one language, however, it must be made clear that these are different languages and one must not mix them up.

A program for multiples of 5.

```
BASIC (generic)        LOGO (LCSI)
10 FOR N = 1 TO 10     TO MULTIPLES5
20 LET M = N * 5       MAKE "N 0
30 PRINT N,M           REPEAT 10 MAKE "N :N + 1
                       [MAKE "M :N * 5 PRINT SENTENCE :N :M]
40 NEXT N              END
```

The children are asked to type the program into the computer, run it, and observe the output and finally use the program and output to explore some relationships. *For example* The following shows examples.

1. (a) The multiples of 5 all end in _?_ or _?_. Why?
(b) Which multiples of 5 end with a 0? What s true about N when M ends with a 0?
(c) What is the difference between any two *consecutive* multiples? (Consecutive means they come one after the other.) Why is this so?

2. Change the program so that the computer prints the first 20 multiples. Are your answers to number 1 still the same?

3. Change your program so that it prints the first 20 multiples of 9.
Run the program.
(a) Is there a pattern in the *units* place? the *tens* place?
(b) Look at the sum of the digits for each multiple. What patterns do
you observe?
(c) What other patterns do you observe in the multiples of 9?

The next task is to consider how one might make the program more
general — that is, to produce as many multiples of a given number
as one wants. The key here is to be able to enter values for the
starting number and the number of multiples, or size of the set. One
can use this setting to teach a bit more of the language, or if the
children have some programming experience they can work on the
task themselves and the focus is totally on the mathematics. In
either event, the result would probably look something like that given
below and again, for your information I have included both BASIC
and LOGO versions of a working program. Note however, that the
children's programs could well be quite different.

A program for multiples of any number

```
BASIC                LOGO
5  INPUT V           TO MULTIPLES :VALUE :SETSIZE
6  INPUT S           MAKE "N 0
10 FOR N = 1 TO S    REPEAT :SETSIZE [MAKE "N :N + 1
20 LET M = N * V          MAKE "M :N * :VALUE PRINT SE :N :M]
30 PRINT N,M         END
40 NEXT N
                     (An alternative using recursion.)
                     TO MULT :V :S
                     IF :S = 0 [STOP]
                     MULT :V :S − 1
                     MAKE "M : S * :V
                     PRINT SE :S :M
                     END
```

Notes
1. Most versions of BASIC also allow words as variables, hence 'V'
could be 'VALUE' or 'value'.
2. For purposes of space the two LOGO programs above use the
short form SE for SENTENCE.

The children can now use the program they have developed, a
procedural description of the idea of a set of multiples, and look for
patterns in sets of multiples, say the multiples of 7, or 11, or 13, or 25.

It can also be used to produce, say, the multiples of any two starting numbers and the children can note any relationships between these sets (a first introduction to the idea of common multiples and 'least common multiple' — for those of you who know a bit about programming, you may like to try to extend the program for multiples to actually print the number of the multiple and both sets in three columns and then look at multiples for say 8 and 12, 7 and 13, 4 and 9 and any others you may like to try. Can you make any generalizations?)

Another point to notice about the programs, either BASIC or LOGO, is that there is a degree of 'transparency' about them — the words used suggest what is happening and the computer output confirms or helps alter one's expectations/thinking and as a result one is learning how the programming words or statements work together. Possibly the most obvious example of this is the FOR and NEXT pair and PRINT in the BASIC programs — of course, this 'transparency' also depends to a certain degree on one's experience, or lack of experience, with a particular language.

Factors
One can support or extend division of whole numbers with computer programming in much the same way as with multiples. An initial short program here might look something like the following (again, BASIC and LOGO versions are given merely to indicate that either could be used.

A program for factors of 18

```
BASIC                 LOGO
10 FOR D = 1 TO 18    TO FACTORS18
20 LET Q = 18/D       MAKE "D 0
30 PRINT D,Q          REPEAT 18 [MAKE "D :D + 1 MAKE "Q
                      18/ :D PRINT SE :D :Q]
40 NEXT D             END
```

The program for factors of 18 can be used to generate output for discussion. As all divisions are output in this program, the children can discuss what there is about each result which will help them decide whether or not a particular number is a factor of 18. While there are some decimal results it is *not* necessary that the children be familiar with decimals in order to make sense of the activity. The decimal can be thought of as a whole number 'and a bit' and a calculator used to find out whether the bit is large or small relative to the divisor (just multiply the whole number part of the quotient, or answer, by the divisor and subtract the result from the original number to get the remainder).

Again, this work can be extended, by the children, to produce alternative, or modified programs and after say two or three refinements they should end up with something similar to that given below.

A program for factors of any number (and only prints factors)
BASIC
```
 5  INPUT N
10 FOR D = 1 TO N
20 LET Q = N/D
30 IF (Q) = INT(Q) THEN PRINT D
40 NEXT D
```

Notes
1. This program uses the greatest integer, INT, function which for positive numbers is only the whole number part of any number — eg, the INT (4.5) is 4.
2. The program could also be made more efficient by changing line 10 to 10 FOR D = 1 TO SQR(N) where SQR(N) is the square root of N, and then printing both D and Q in line 30.

LOGO
```
TO FACTORS :NUMBER
MAKE "D 0
REPEAT :NUMBER [MAKE "D :D + 1 IF 0 = REMAINDER
:NUMBER :D [PRINT :D]]
END
```

Notes
1. It is not necessary to use INT in the LOGO program as one can use the idea of a REMAINDER and REMAINDER 16 5 gives the remainder for 16 ÷ 5 or 1. If one number is a factor of another, the remainder is 0 and this fact provides the basis for the IF statement in the program.
2. One could have used INT in the LOGO program — and in this case the REPEAT line would look something like
REPEAT :NUMBER [MAKE "D :D + 1 IF :NUMBER/:D = INT
:NUMBER/:D [PRINT :D]]

This procedural description of the idea of the factors of any number can now be used to look at numbers and find some with many factors, or with few, say two, or three factors — what kind of numbers are these? Of course, a next step might be to think about what one might have to do to extend or change the program to test whether or not a number is a prime number. Remember our friend the prime number? And from here how about looking for 'perfect numbers' (see Activities 19-25 in Johnson, 1983, pp44-58).

Summary
Points A1 and A4 - mathematics as dynamic procedures in a formal system
While the main emphasis in the examples in the previous section has been on the potential of (dynamic) procedural descriptions of concepts and ideas in the teaching and learning of mathematics you should have also noticed the importance of being precise in the setting out of the procedures. I can share with you the fact that many of the sample programs included here were not first attempts — I needed to do some experimentation as well as reflect on what I wanted the computer to do and then the preparation of the final program required both 'precision and rigour', an aspect of 'mathematics as a formal system'.

Point B1 — learning to program
Another feature of the examples used in multiples and factors was that of the role of 'blocks of code' or short programs for learning about programming as well as mathematics. Of course, as the children become more experienced the role of the short programs becomes more one of 'getting started' in the study of an idea or topic and they take on a lesser role in the teaching of the language itself. Reference manuals and books will provide the resource for answering questions about the syntax and semantics of the language (sorry about the jargon).

Point B3 — the question of language
We now return to the question of which language: or really to why I feel this is not *the* key question. I have attempted to present a case here for using any language which satisfies some criterion for acceptability and is also readily available. To be acceptable the language should have suitable words, structures (eg, loops and conditional, the IF) and the 'working blocks of code' or short programs should not appear as collections of 'disjoint information', ie, they should make sense. Further, the choice of language is unimportant as long as the language chosen is one which facilitates communication and any idiosyncrasies, ie, the special words, characters, punctuation, ways of doing things, etc, do not get in the way of working on the central task, in this case, doing mathematics. The 'bottom line' here is that 'BASIC is perfectly acceptable for

doing mathematical programming'. And, of course, so are LOGO and COMAL (and others). On the other hand you must consider 'what mathematics' and with 'what age group'. As an aside, however, it is not only a matter of choosing a language, but also how one teaches it. One can abuse even LOGO, the 'language for young children', by attempting to present advanced features before they are seen to be needed.

As is suggested by the last comment in the preceding paragraph, how one goes about introducing and teaching the computer language, regardless of how good or bad some people think the language is, is crucial. We have *some limited* evidence as to how best to proceed here (see the introduction to this paper), and *substantial* evidence as to how *not* to do it. It is clear that a programming language should *not* be taught as a set of isolated words and structures or from a building up of a 'word at a time', but rather words and structures should be seen in a context of working together to do significant and/or interesting tasks. Recent work in the UK (Noss, 1985) has much to say here, particularly in terms of a model for learning the computer language LOGO.

Epilogue — a caution
There are many other examples of concepts as procedures which could have been used in this paper. A classic LOGO turtle example is that of developing a procedure for a circle (a pseudo-circle that is) — move a little, turn a little, move a little, turn a little, etc. On the other hand, think about the question of 'how does this help me to draw a circle with a radius of say 50 and whose centre is the centre of the screen (or paper)?' Further, try to draw an ellipse with a move a little, turn a little type procedure. An ellipse is the shape formed when you draw the curve such that the sum of the distances from two fixed points is constant. You can draw this nicely using a piece of string which is longer than the distance between the two points. Attach the string to the two points and move your pencil around the points keeping the string taut. But of course this relies mainly on the definition, not any procedural description (or does it?). What this says is that the case can be made for more than one view of a mathematical idea or relationship. The question of 'which is best' is much like that of 'which programming language is best' — we don't know, and the answer quite likely depends on 'for what purpose'. This is not meant to suggest we can delay on implementing a procedural approach but quite the reverse.

The inclusion of an opportunity for children to study and explore mathematics through programming procedures will make the subject come alive — as mathematics really is dynamic.

References

JOHNSON, DC, *Explore Maths with your Micro: a book for kids aged 9 to 90,* Heinemann, London, 1983

NOSS, R, *Creating a Mathematical Environment through Programming: a study of young children learning LOGO,* University of London, 1985

Starting mathematical graphics on the microcomputer

Trevor Fletcher, HMI

Getting started

Do you 'do it yourself' or do you seek out software that other people have written? This question will always face any user of microcomputers who wishes to start out in an unfamiliar area. When we are concerned with mathematical graphics both strategies are available and there is no reason why both should not be followed at the same time. Many good graph-drawing programs are readily available and it is useful to have some of them to hand in a mathematics room. Professionally written packages or programs taken from hobbies magazines produce block graphs and piecharts with an extra degree of polish, and they should certainly be used when appropriate. But graphics writing of this kind is not difficult to do and there is a strong case for including some of it in the standard mathematics courses for most pupils. The strongest initial justification for coordinate geometry nowadays is surely that it enables you to do geometry on machines. Many pupils can produce precise, immaculate work on a microcomputer which is far beyond anything which they could do by hand. This article is written for those who wish to introduce 'do it yourself' graphics into the mathematics classroom.

It is not possible by the printed word alone to explain all of the little points of detail which you need to know in order to progress on a microcomputer without irritating holdups. You need to know how to manage the keyboard, how to move the cursor about and how to reduce work by using the 'Copy' key. If you are not familiar with these matters it is best to consult someone who is and have 10 or 20 minutes tuition at the keyboard. This article has been written to help teachers who have this minimal knowledge to get started with their own graphics and to pass some of this skill on to their pupils.

It is a handicap that the conventions and the details of programming differ on different machines, and this is particularly true of the graphics commands. Because of this it is usually necessary to have a particular machine in mind when discussing graphics. This article is written in terms of the BBC machine, but to help those who use other machines many of the programs are kept as simple as possible and some of the individual features of BBC BASIC are not exploited.

The early stages of computer graphics could hardly be simpler. You need only two commands — MOVE and DRAW — and these mean just what they say. To use these commands you have to know what you 'move' and 'draw' on; and you draw on the grid which is shown on page 495 of the manual(2). This means that you use coordinate geometry with x in a range from 0 to 1279 and y in a range from 0 to 1023. Anyone who knows how to describe a point on a grid by its coordinates can make a start, and you pick up the rest as you go along.

Not all of the eight modes on the BBC micro provide graphics. When the machine is switched on it starts up in mode 7 which provides only a special kind of graphics of its own. So before you can use any of what follows you must switch to a graphics mode. Type either MODE 1 or MODE 4 and follow it by pressing the RETURN key. (The RETURN key has to be pressed at the end of every line in a program or at the end of every instruction if you are commanding the machine directly.) At the present stage it does not matter at all whether you use mode 1 or mode 4; mode 1 provides four colours instead of two but this requires more storage space inside the computer. This does not matter until you start to write large programs.

To begin with you may type in anything you like of the form

```
DRAW 400,500
DRAW 700,300
MOVE 100,100
DRAW 900,800
```

You should quickly find that these instructions do what they say. We are controlling the movement of a pen, and if we say 'draw' it draws to the point named, and if we say 'move' it moves without leaving a trace. If at any time you want to run everything out and start again type CLS and follow it, as usual, with RETURN.

A great deal can be done with no more than this, and the best course of action will depend greatly on the circumstances in the classroom — on the ages of the pupils, on their previous knowledge of geometry, on the number of computers which are available, and on the way in which the school has chosen to deploy them.

Some teachers have started by prescribing simple tasks, such as drawing squares or rectangles in various parts of the screen; others have given the pupils an almost free rein. There is much to be said at the beginning for allowing the maximum of free experiment. Many pupils would sooner follow their own hunches than execute set tasks and they certainly learn at least as much at this stage. If in the early stages the pupil is asked merely to 'draw something' it is difficult to be wrong! Pupils can fail on the simplest prescribed tasks, and if the task is for everyone to draw a square then mistakes are clearly displayed, especially if the pupils do not yet know how to erase lines.

Pupils' response to freedom in such matters varies, particularly if they have not been used to it before. The best strategy is perhaps to have a selection of prescribed tasks of increasing difficulty available, and to allow individual initiative to take its course when this seems reasonable. There is no doubt that, at any stage, pupils who are deciding their own objectives are willing to draw figures of far greater complexity than the teacher would dare set as a prescribed task.

When you are ready to introduce some more technicalities make sure that you are in mode 1 (and not mode 4) and type GCOL0,2. You should then find that subsequent lines are drawn in colour. GCOL obviously stands for 'graphics colour', the zero indicates the type of drawing (more about this later), and the two indicates the colour. In mode 1 you can use up to four colours at any one time. Colour 3 is the white that you get if you do not specify any special requirement. Colours 1 and 2 are red and yellow respectively, and Colour 0 is black. So now we see how to rub out individual lines — go over them again in black. (Try it.) A range of different colours can be obtained by using the VDU19 command (for this see the manual).

With very little technical knowledge you can quickly produce very effective diagrams of the kind which are often produced in the classroom by curve stitching. One such diagram is produced by marking out equally spaced scales on a pair of lines at right angles and numbering them from the point of intersection. The points may be numbered (say) from 0 to 20, starting at the corner. Then stitch (or draw) from point 0 on the first scale to point 20 on the second, then point 1 to point 19, point 2 to point 18 and so on.

How do we do this on a microcomputer? Looking at the sizes involved we see that it might be a good idea to go up to 800 by steps of 20. So our first tentative program is

```
10   MODE1
20   MOVE 0,0
30   DRAW 0,800
40   MOVE 20,0
50   DRAW 0,780
60   MOVE 40,0
70   DRAW 0,760
```

and so on. But we quickly tire of this — there must be a better way, and this is a suitable moment at which to introduce the idea of a

variable. We can explain the idea of a FOR loop. After suitable
discussion we can arrive at the solution

```
10   MODE1
20   N = 0
30   MOVE N,0
40   DRAW 0,800 – N
50   N = N + 20
60   GOTO30
```

This is an imperfect temporary solution; but try it and you will
find that you get something like the drawing you want. GOTO is
an instruction that is self-explanatory, and whilst you should
quickly move on to rather different methods GOTOs enable
beginners to get started.

It is a more serious objection to this program that it never
stops! If the machine gets into an endless loop with this kind of
program then press the ESCAPE key and normal working is
restored. One solution is to alter line 60 to IF N < 800 THEN
GOTO20. (THEN is not needed in BBC BASIC, but it is needed on
some other machines.) But it would be better to reorganize this
rough-and-ready program, and produce something such as

```
10   MODE1
20   FOR N = 0 TO 800 STEP 20
30   MOVE N,0
40   DRAW 0,800 – N
50   NEXT N
60   END
```

Whether to include a systematic course in programming in the
pupils' curriculum is a separate issue. A few very simple and
self-explanatory instructions are sufficient to produce rewarding
drawings which well repay the short time needed to get started.

As you produce graphics on a microcomputer you find that
there is an interplay between the ideas you are trying to express
and the techniques you need to make the machine operate. This
offers many teaching opportunities. There are nearly always
problems with scaling diagrams and with arranging for the output
of the computer to be in a suitable form. But scaling and
formating are not undesirable chores, they are worthwhile
mathematical activities with lessons to teach.

To take a simple case consider the problem of drawing $y = x^3 - 3x$ between $x = -2$ and $x = 2$. In the computer program the algebraic expression will have to be written as $y = x * x * x - 3 * x$.

We put the origin in the centre of the screen by means of VDU29,640;512;. We see that when $x = 2$ the value on the screen will need to be about 500 or 600, which indicates a scaling up by a factor of 250 or 300. A rough calculation shows that y also needs to be between -2 and $+2$, and since the screen range goes up to 512 maximum (now that the origin is in the centre) a scale factor of 250 should do for both x and y. This process is no more difficult than the scaling which usually has to be undertaken to fit a graph on to a given piece of graph paper. We can work out that our graph is going to start at $x = -12$, $y = -2$ and a first shot at a program might take the form

```
10   MODE1
20   VDU29,640;512;
30   scfactor = 250
40   MOVE - 2 * scfactor, - 2 * scfactor
50   FOR x = - 2 TO 2 STEP 4/100
60   DRAW x * scfactor,(x * x * x - 3 * x) * scfactor
70   NEXT
```

This is not a very expert program but it gets you started. When it is running satisfactorily experiment by altering some of the details. We put in a step of 4/100 because we know that the range is 4 and 100 steps seems a good number to try first. Note that we do not even have to divide 4 by 100 ourselves, and indeed it is better not to, because if you want to try using a different number of steps it is immediately clear which number has to be changed. Once you have started, with a good instruction book on the practical details of elementary programming, you quickly get ideas of your own. The graph so far does not have axes or scales. If you want these try to add lines to the program which draw them in.

Contours

Drawing graphs is a different matter if y is not given explicitly as a function of x. To draw the graph of $x^3 + y^3 - 3xy = 0$ is, by previous standards, quite a difficult task. But with a microcomputer fresh methods are available. We can outline the curve by drawing the two sides of the graph in different colours! To do this visit each point of the screen in turn, evaluate the algebraic expression, DRAW if the expression is positive and pass on if it is not. A suitable program is given below, but a few technical points have to be clarified. The scaling is, as with ordinary graph drawing, a matter of judgement and experiment — but it is much easier with a computer to adjust your guesses and try again. Whilst the coordinate grid on the computer screen is as we described above, the screen having 1280×1024 addressable points, the points are clustered in groups to enable plotting to be in different colours. These groups are called 'pixels', and in mode 1 the pixels are four points by four points in each direction. For this reason we do not proceed by steps of one in the program but by steps of four. The command PLOT69 is simply the command to plot a point. (Again, see the manual for further explanations.)

The program also makes use of variables followed by the percentage sign, such as $X\%$. This has nothing to do with the usual meaning of the sign; in BBC BASIC the percentage sign indicates that the variable is to be handled as an integer. It is good to use integer variables whenever you can, especially when speed is important, and the variables $A\%$ to $Z\%$ are faster than integer variables with names such as coord%.

```
10   xmax = 2.1
20   s = xmax/640
30   MODE1:VDU29,640;512;
40   FOR X% = −640 TO 640 STEP4:x = X% ∗ s
50   FOR Y% = −512 TO 512 STEP4:y = Y% ∗ s
60   IF x ∗ x ∗ x + y ∗ y ∗ y − 3 ∗ x ∗ y > 0 PLOT69,X%,Y%
70   NEXT:NEXT
80   END
```

This program takes more than half an hour to run — so be prepared! The diagram is very effective, but even more attractive diagrams can be produced by taking a little more care with the drawing instructions. When plotting algebraic functions in this way we are effectively dealing with a quantity z (say) which can be seen as a function of x and y. What we are doing is plotting the contour line where $z = 0$. But with a computer it is just as easy to plot a whole set of contour lines as it is to plot the one. Furthermore, as with a geographical map, we can colour the diagram to indicate the 'height'. This leads us to modify the program by replacing line 60 by lines such as

```
60   C% = INT(4 ∗ (x ∗ x ∗ x + y ∗ y ∗ y − 3 ∗ x ∗ y))
62   C% = (C% + 200) MOD4
64   GCOL0,C%: PLOT69,X%,Y%
```

The factor 4 in line 60 is to produce a suitable range of variation in the function; it amounts to scaling the height. If you do not already know you need to investigate how the commands INT and MOD behave on the computer. They are used here to transform the value of the function into 0,1,2 or 3 to use in the colour command. Drawing like this calls for a certain willingness to experiment, and different stratagems have to be adopted in different circumstances; but the results amply reward the time spent.

The knot curve, with equation $(x^2 - 1)^2 = y^2(3 + 2y)$ is a good one to try, and a range of further suggestions may be found in Cundy and Rollett(4). With a suitable algebraic expression impressive drawings can sometimes be produced by taking the sine of the whole expression and drawing in black or white as the sine is positive or negative. Some expressions get out of hand because of the wide range of numerical values which they take. In such cases it often helps to use the ATN (inverse tangent) function. This compresses an infinite range of values on to the finite interval between plus and minus pi. (Remember that the trigonometric functions on a micro use radians.) If we now use DEG this range goes between plus and minus 180, and this can easily be coloured by our previous methods.

As a further example we may consider the problem of drawing bargraphs to illustrate binomial distributions. (Pass over this example if you think it is too difficult.) Bargraphs can most simply be drawn by moving to a point and drawing in a vertical line of the appropriate length. The program which follows is a little more elaborate as it draws rectangles and fills them in. The fill command is PLOT85 (look this up in the manual). This example contains further refinements which you may wish to omit if you are still at the early stages, but they illustrate further ideas which you can incorporate in your programs as you progress. The program provides prompts on the screen so that the user can run the program several times with different input. The methods of doing this are not difficult to learn, and attention to detail of this kind produces programs which are more convenient for classroom demonstration and better for other people to use.

Here then is a block graph to show binomial distributions. The binomial distribution gives the expected outcome when numbers of coins or dice are tossed, and the theory can be found in O-level or A-level books on statistics. The width of the rectangles is varied according to the number to be drawn. The height of the rectangles is harder to decide, and the program employs a scaling (discovered by experiment) which is satisfactory in most cases; but the user is given the option of rescaling if this helps.

```
5 REM By TJF, October 1982
10 MODE7
20 PRINT:PRINT:PRINT:PRINTTAB(10) "BINOMIAL BAR GRAPHS"
30 PRINTTAB(10) " = = = = = = = = = === ======="
40 PRINT ' ' "This program draws binomial bar graphs using input
   parameters which you supply"
50 PRINT ' ' "After the graph has been drawn you may press any
   key to continue"
60 PRINT ' ' "(NOTE: with large values of n it is more accurate to
   work with small p; so use 1 – p rather than p if p > 0.5 and n is
   big)"
70 INPUT ' ' "WHAT IS p",p
80 INPUT "WHAT IS n",n
90 B = 20
100 vscale = 2500
110 PRINT "The vertical scale factor is ";vscale
120 INPUT "Do you want to alter this (Y/N)",a$
130 IF a$ = "N" GOTO 140 ELSE INPUT "What is the vertical scale
    factor",vscale
140 W = 1280/(n + 4)
150 CLS:MODE0
160 FOR r = 0 TO n
170 IF r = 0 THEN y = (1 – p)^n ELSE y = y * (n – r + 1) * p/(r * (1 – p))
180 YS = y * vscale
190 IF YS < 4 THEN YS = 4
200 MOVE (r + 2) * W,B
210 DRAW (r + 2.9) * W,B
220 PLOT 85,(r + 2.9) * W,B + YS
230 PLOT 85,(r + 2) * W,B + YS
240 PLOT 85,(r + 2) * W,B
250 NEXT r
260 A$ = GET$
270 RUN
280 END
```

The interplay of mathematical ideas and graphics techniques is
further illustrated by the use of matrix transformations. In some
ways matrix ideas may have been introduced into syllabuses before
their time and with insufficient regard for applications; but nowadays
matrices find immediate application in microcomputer graphics. To
make changes in your drawings as often as not a matrix
transformation is what you need. This is especially true with
rotations and with three-dimensional drawing. Although some
excellent ready-made programs are available to illustrate the work

on matrix transformations which occurs in many O-level and CSE syllabuses it is instructive to build up a package for yourself, introducing the various transformations step by step. Ideas for this may be found in Oldknow and Smith(7).

Drawing circles

Sooner or later you want to draw circles. Microcomputers never actually draw circles, they draw polygons with so many sides that you cannot see the difference. When we speak of 'circles' we mean circles in this sense. Circles come in to LOGO graphics in a way which calls for none of the usual school theory; but LOGO has its own very individual approach to geometry and there is not space to consider it here.

Circles can be drawn by a variety of methods. It is possible to write a circle-drawing program once the theorem of Pythagoras is understood, but the most straightforward way is to use the sine and cosine functions. The introduction of sine and cosine in the secondary course has never been an easy matter. The traditional method was to introduce these functions as ratios. Many pupils are uneasy with ratios, and they are uneasy afresh each time some new ratio is introduced. The projects of the 1960s introduced cosine and sine as the coordinates of points on a circle of unit radius. In this way sine and cosine are simply lengths, which are easier to understand than ratios, and there is the added advantage that sine and cosine are defined for all angles and not merely for angles between 0 and 90 degrees. This approach is not entirely free of ratio, because to use the trigonometric functions with triangles of different sizes the radius of the initial circle has to be scaled approximately, and scaling involves ratio. However, the foundations of this can be laid with care earlier in the course. Drawing a diagram to scale is a natural idea whose purpose it is easy to understand — and so when scaling up is needed in trigonometry it is an idea which has already become familiar in a previous context.

Microcomputer graphics encourage this approach. The COS and SIN commands in the language are there to work out the coordinates of points on a circle. We need to work with the origin in the centre of the screen. BBC BASIC has a special command to arrange this: VDU29,640;512; (take care with the semicolons!) The rationale of this command is explained in the manual (page 388). After this, the point on a circle of unit radius with a bearing which we can call 'angle' has coordinates COSangle and SINangle. (Note that in programming it is often the habit to use words rather than symbolic abbreviations; this makes programs easier to read if you come back to them after some time.) If the circle has radius R then the coordinates have to be multiplied by R. (Program using the word 'radius' if you wish, although R% would be quicker.)

There is one snag to this approach. The computer prefers to measure angles in radians rather than degrees! This is unfortunate in the context of normal mathematics teaching and a short time has to be devoted to overcoming the difficulty. If you are to understand how to draw circles on a micro you have to know how the micro expects angles to be measured. Angles are measured in radians by, as it were, taking a protractor of unit radius and measuring the angle by the length of the corresponding arc. Hence 180 degrees is pi radians, 90 degrees is pi/2, and so on. It is not necessary to practise conversions between degree and radian measure because the machine will do this. The commands RAD and DEG convert for you. (Type in such commands as PRINT RAD180, PRINT DEG3.14 and check this.) With these commands you may put the angle in brackets if you wish, it does not make any difference.

Therefore if we are working in degrees and we wish to get the coordinates of points on a circle with centre at the origin, the appropriate instructions are COSRADangle, SINRADangle. It is not stretching the truth too far to regard COSRAD and SINRAD as the appropriate words for cos and sin if you are using a micro — but it goes without saying that a teacher needs to provide a fuller explanation when it is called for. So we can draw a circle with

```
10  MODE1
20  VDU29,640;512;
30  radius = 400
40  MOVE radius,0
50  FOR angle = 0 TO 360 STEP10
60  DRAW radius * COSRADangle,radius * SINRADangle
70  NEXTangle
80  END
```

Once this is available the possibilities are immense. As a variation on the previous curve-stitching program try

```
10  MODE1
20  VDU29,640;512;
30  r = 500
40  FOR angle = 0 TO 360 STEP 5
50  MOVE 0,r * COSRADangle
60  DRAW r * SINRADangle,0
70  NEXT angle
80  END
```

Another curve-stitching favourite is produced by joining the points 'angle', 'twice-angle' by a thread. This can be drawn with

```
10  MODE1
20  VDU29,640;512;
30  r = 500
40  FOR angle = 0 TO 360 STEP 5
50  MOVE r * COSRADangle,r * SINRADangle
60  DRAW r * COSRAD(2 * angle),r * SINRAD(2 * angle)
70  NEXT angle
80  END
```

Lissajou figures make most effective diagrams and are very easy to draw. The essential idea is to take the x and y coordinates as trigonometric functions of a variable which we may continue to call 'angle'. The essential line within your loop then becomes something like

DRAW 500 * COSRAD(angle + 30),500 * SINRAD(2 * angle)

You may experiment with a great variety of similar expressions. Science textbooks are more likely than mathematics books to tell you about Lissajou figures.

Another popular curve-stitching diagram is sometimes called the 'mystic rose'. This is produced by placing a number of points equally spaced around a circle, and joining up each one to all of the others. You can include in your program a line such as

INPUT N

When the program runs it will pause at this stage and wait for you to type in a number (and RETURN). The next line of the program could be

angle = 360/N

This gets us started on writing a mystic rose program, and we leave the rest as a problem for the reader. The essential idea is to move to each of the N points in turn and draw to all of the others. (This is wasteful, but perhaps you can improve on the idea.) You need a pair of loops. Your program will have to contain lines such as

```
FOR A = 0 TO N
FOR B = 0 TO N
. . . .
. . . .
NEXT B
NEXT A
```

Note that with BBC BASIC you can make programs easier to read (but not to type) by using long variable names. Thus in the above outline instead of using A we could use a name like 'startpoint', and we could call B 'endpoint'. It is a good idea to cultivate habits like this (where speed is not a major concern) because programs can be made to look more like ordinary English; in this way they are less offputting to those who do not see themselves as mathematicians.

Many of the diagrams in Lockwood(6) are very effective when drawn on a computer.

Moving pictures

Microcomputer graphics are not restricted to static diagrams; movement is possible, and this movement is an essential component of many popular games. Moving diagrams are useful in teaching mathematics, especially in showing how geometrical figures change as a parameter is varied.

The scope for simple movement is greatly increased by drawing out as well as drawing in. We will take an illustration from mechanics, at sixth-form level on the borderland of mathematics and physics, which shows a movement which is not easy to visualize although it can be demonstrated in general terms with simple physical apparatus. Three particles on an elastic string execute coupled oscillations. The following microcomputer simulation shows only the displacements of the particles as they vibrate, but it provides a demonstration in which it is possible to control the various parameters and investigate the effect of changing them. The program also shows how finite difference methods provide a mathematically simpler approach to physical problems then differential equations. There is not space here to discuss the mechanical theory involved, but we may note that arcade games use finite difference methods to control moving objects, and that the controls are of two types — some change the position of the moving object and some change its velocity. The 'feel' of the two types of control is quite different and many students readily adapt to the two situations even though they may find the initial ideas of Newtonian mechanics very difficult. But the difference is precisely the difference between Newtonian mechanics and what went before. Newton showed that a force produces acceleration rather than velocity. The techniques used in the programming of arcade games can be used to teach mechanics.

The example which follows may not be very interesting by arcade standards but it will serve to introduce draw-in draw-out techniques. A number of remarks in the program indicate the strategies employed. When programs get to this size it is an excellent idea to break them up in procedures. They are much easier to read and to modify if this is done, and the procedures can frequently be used again in other programs. Procedures are not used here because we are deliberately trying to limit the number of programming ideas which are employed, and to write programs which transfer without too much difficulty to other machines.

```
10 REM MODES OF VIBRATION OF THREE PARTICLES
20 MODE7
30 PRINT' "VIBRATIONS OF THREE CONNECTED PARTICLES"
40 PRINT "= = = = = = = = = = = = = = = = = = = = = = = = = = ="
50 PRINT' ' "The program displays the oscillations of three
   particles on a string."
60 PRINT' "The initial positions may be set."
70 PRINT' "The other parameters may be changed by modifying
   the program."
80 PRINT' "The initial position is displayed; to continue press any
   key."
90 PRINT' "Initially keep displacements below 2."
100 INPUT' ' "Displacement of first particle",X
110 INPUT ' "Displacement of second particle",Y
120 INPUT ' "Displacement of third particle",Z
130 REM Following constants are force constants given by the
    mechanical theory
140 A = 2:B = -1:C = 0:D = -1:E = 2:F = -1:L = 0:M = -1:N = 2
150 REM Get into drawing mode and specify drawing parameters
    etc
160 MODE5:VDU29,640;512;
170 VDU23;8202;0;0;0;
180 H = .05:S = 200
190 REM Make initial velocities zero: draw in initial positions
200 U = 0:V = 0:W = 0
210 MOVE -250,0:DRAW -250,S * X
220 MOVE 0,0:DRAW 0,S * Y
230 MOVE 250,0:DRAW 250,S * Z
240 REM Set 'old' values of X,Y,Z, equal to present values
250 P = X:Q = Y:R = Z
260 A$ = GET$
270 REM Animation cycle follows
280 REPEAT
290 U = U + H * (A * X + B * Y + C * Z)
300 V = V + H * (D * X + E * Y + F * Z)
310 W = W + H * (L * X + M * Y + N * Z)
```

```
320 X = X - H * U:Y = Y - H * V:Z = Z - H * W
330 MOVE  - 250,S * P: IF X * (X - P) > = 0 PLOT 5, - 250,S * X
    ELSE PLOT 7, - 250,S * X
340 MOVE 0,S * Q: IF Y * (Y - Q) > = 0 PLOT 5,0,S * Y ELSE PLOT
    7,0,S * Y
350 MOVE 250,S * R: IF Z * (Z - R) > = 0 PLOT 5,250,S * Z ELSE
    PLOT 7,250,S * Z
360 P = X:Q = Y:R = Z
370 UNTIL FALSE
380 END
```

If you wish to add movement to conventional geometrical diagrams a little more programming technique is needed. The first thing required is a knowledge of EOR plotting. The GCOL command in BBC BASIC enables you not simply to draw, but to draw in such a way that the colour resulting is a logical combination of the colour which was there previously and the colour which you apply. Three important notions are involved: the logical notions of AND, OR (or both) and EOR (exclusive or, that is or-but-not-both). These notions relate logical ideas in precisely the same way that intersection, union and disjoint union relate 'sets' in Venn diagrams. The ideas were introduced into school mathematics with the reforms of the 1960s, but they were often introduced without regard to their applications with the result that they are still comparatively little appreciated.

The logical commands act on the colour numbers 'bitwise', that is one bit at a time, If, in a four-colour mode, the existing colour at some point on the screen is colour 3 (binary 11) and we apply colour 2 (binary 10) with the EOR command, GCOL3,2, then the resultant colour is 1 — because if we take 11 and 10 and 'EOR' the bits in order we get 01 (which is the binary number of the resultant colour). If we now apply colour 2 a second time with the EOR command we get the colour 01 EOR-ed with 10 again, and this time the resultant is binary 11, or 3. The important thing is that EOR-ing twice always restores the status quo. Much of the movement of the characters in arcade games is produced by EOR drawing.

In passing we may observe that the BBC micro is very willing indeed to do bitwise logic on integers, because it is useful in various aspects of computing. Using the machine in command mode experiment with instructions such as

PRINT 11 AND 7
PRINT 12 OR 5
PRINT 9 EOR 10

Here is a first attempt at animating a diagram showing the
constancy of the angle in a given segment. We compile a look-up
table of sines and cosines so that the program runs more quickly
during the animation. At line 290 we actually introduce a little delay
in order to hold the picture on the screen for a short while. The
amount of delay can be controlled by varying line 30.

```
 10 REM ANGLE IN SEGMENT
 20 REM By TJF January 1985, uses EOR drawing
 30 delay = 5
 40 A% = 150:B% = 105
 45 REM complete look-up table
 50 DIM S(180),C(180)
 60 PRINT' ' "WAIT"
 70 FOR N% = 0 TO 180
 80 S(N%) = 400 * SIN(RAD(2 * N%)): C(N%) = 400 * COS
    (RAD(2 * N%))
 90 NEXT
100 MODE1
110 VDU29,640;512;
120 VDU19,3,2,0,0,0
125 REM draw circle
130 GCOL0,1
140 MOVE C(0),S(0)
150 FOR N% = 1 TO 180
160 DRAW C(N%),S(N%)
170 NEXT
175 REM draw angle
180 GCOL3,2
190 MOVE C(A%),S(A%)
200 DRAW C(0),S(0)
210 DRAW C(B%),S(B%)
215 REM animation loop follows
220 REPEAT
230 FOR N% = 1 TO 180
240 M% = N% - 1
250 DRAW C(M%),S(M%)
260 DRAW C(A%),S(A%)
270 DRAW C(N%),S(N%)
280 DRAW C(B%),S(B%)
290 TIME = 0:REPEAT: UNTIL TIME > delay
300 NEXT
310 UNTIL FALSE
```

When you run this program you will, let us hope, be delighted that you have got some sort of movement into the diagram, but you may not like the flicker which is present. The flicker can be reduced by taking more trouble with the drawing and 'switching the colours'. Colour switching is described in such references as Cownie(3). The essential point is that we always draw with logical colours (in a four-colour mode this means with colours labelled 0 to 3); but the actual colours which correspond can be changed with a further command to the machine, VDU19; and these changes can be made very quickly. Therefore it is possible to draw each new frame of an animation in black and switch it to white when it is complete. Simultaneously you switch out the old frame, and you then delete it with a second EOR command whilst its actual colour is black. This is difficult to explain by word alone, but the details of the program should be studied with care because they are the key to much mathematical animation.

```
 10 REM ANGLE IN SEGMENT
 20 REM By TJ Fletcher April 1985, uses EOR drawing with colour
    switching
 30 delay = 5
 40 A% = 150:B% = 105
 45 REM Compute look-up table
 50 DIM S(181),C(181)
 60 PRINT' ' ''WAIT''
 70 FOR N% = 0 TO 181
 80 S(N%) = 400 * SIN(RAD(2 * N%)): C(N%) = 400 * COS
    (RAD(2 * N%))
 90 NEXT
100 MODE1
110 VDU29,640;512;
120 VDU19,3,1,0,0,0
130 GCOL0,3
140 MOVE C(0),S(0)
150 FOR N% = 1 TO 180
160 DRAW C(N%),S(N%)
170 NEXT
180 VDU19,1,0,0,0,0,19,2,3,0,0,0
190 GCOL3,2
200 MOVE C(A%),S(A%)
210 DRAW C(0),S(0)
220 DRAW C(B%),S(B%)
230 GCOL3,1
```

```
240  DRAW C(1),S(1)
250  DRAW C(A%),S(A%)
260  VDU19,1,3,0,0,0,19,2,0,0,0,0
265  REM Main animation loop follows
270  REPEAT
280  FOR N% = 0 TO 178 STEP2
290  GCOL3,2
300  DRAW C(N%),S(N%)
310  DRAW C(B%),S(B%)
320  DRAW C(N% + 2),S(N% + 2)
330  DRAW C(A%),S(A%)
340  VDU19,1,0,0,0,0,19,2,3,0,0,0
345  TIME = 0:REPEAT: UNTIL TIME > delay
350  GCOL3,1
360  DRAW C(N% + 1),S(N% + 1)
370  DRAW C(B%),S(B%)
380  DRAW C(N% + 3),S(N% + 3)
390  DRAW C(A%),S(A%)
400  VDU19,1,3,0,0,0,19,2,0,0,0,0
410  TIME = 0:REPEAT: UNTIL TIME > delay
420  NEXT
430  UNTIL FALSE
```

The method of EOR plotting enables you to do a lot of animation but difficulties arise if parts of the drawing have, for some reason, to be drawn in twice for some frames of animation. This happens, for example, in the drawing of a tumbling box which is given in the set of 'Creative Graphics' programs(3) from Acornsoft. In these circumstances OR-plotting (using GCOL1) and AND-plotting (using GCOL2) are required.

If you wish to follow up further methods of producing effective mathematical diagrams sources such as (3), the books by Kosniowski(4) and by McGregor and Watt(8), and the hobbies magazines on computing have every bit as much to offer as the specifically educational publications. The microcomputer offers a fresh start in mathematics in many different ways. The appeal of graphics can do much to vitalize teaching — especially as teachers and pupils can produce novel, exciting and highly instructive graphics for themselves without having to depend on specially produced packages.

References
1. ANGELL, IO and JONES, BJ, *Advanced Graphics with the BBC Model B Microcomputer,* Cambridge University Press, 1983

2. COLL, J, *The BBC Microcomputer User Guide,* BBC, 1982

3. COWNIE, J, *Creative Graphics on the BBC Microcomputer* (book and tape), Acornsoft, 1982

4. CUNDY, H M and ROLLETT, A P, *Mathematical Models,* Cambridge University Press, 1952

5. KOSNIOWSKI, C, *Fun Mathematics on Your Microcomputer,* Cambridge University Press, 1983

6. LOCKWOOD, E H, *Curves,* Cambridge University Press, 1961

7. OLDKNOW, A J and SMITH, D V, *Learning Mathematics with Micros,* Ellis Horwood, 1983

8. McGREGOR, J and WATT, A, *The Art of Microcomputer Graphics,* Addison-Wesley, 1984

Mathematical imagery and computers

Seamus Dunn, Education Centre, New University of Ulster, Coleraine, Northern Ireland

'This personal symbolism is particularly alive in mathematicians themselves. Like all creative people they have continual recourse to their imaginations, whether they think as analysts or geometers. The minds which are best able to control abstractions are those which succeed in embodying them in concrete examples or schemes which then serve as symbolic springboards without introducing any limitations' (Piaget and Inhelder, 1971, p10).

Mental imagery

Most textbooks and pupil materials about school mathematics include a built-in, perhaps unexamined, view that a picture or a diagram or a figure is a useful aid when presenting mathematical ideas. This pictorial approach is for many teachers instinctive: when there is a need to demonstrate an idea, solve a problem or explain a difficulty, a diagram related to the situation seems an obvious help. Indeed much of what passes for problem-solving in school mathematics is about the manipulation of diagrams which represent a model (or an image or a metaphor) for the issues involved. A picture is seen as a concrete representation of an abstract idea.

The importance of this issue is of course now highlighted by the emergence of the microcomputer, since its capacity for communication using pictorial imagery is, at least potentially, one of its most innovative characteristics. Most computer-assisted learning packages now being produced involve the use of pictorial imagery in some form, and the whole world of turtle geometry rests on a graphical presentation of ideas. There is a widespread view that many mathematical ideas held to be difficult and 'abstract' can be transmitted and understood if the medium of transmission is pictorial. As often as not this view is implicit, but it can also be quite explicit. The quotation from Piaget and Inhelder above is an example of this, and the writings of Gattegno provide a powerfully held personal version of this view: '. . . the stuff of geometry was the mental stuff called images . . .' (1965, p22).

The purpose of this article is to examine this view and to demonstrate in a small way some results with children. This involved an experiment using a BBC microcomputer, some programs, and some groups of children. The intention was to examine the effects of allowing children to view the equivalent of an animated film, with the added possibility of some limited interaction. It was hoped that this would result in a number of hypotheses related to the issue of mental imagery, learning and mathematics.

Much of what is believed about the use of pictures is uncritical in the sense that no examination is made of the hidden assumptions involved, the most pertinent of which is a belief in the importance and the value of mental imagery, especially mental imagery of a pictorial nature. For this reason it is useful and salutary to refer to the psychological literature on the subject which is large, complex and full of unanswered questions and issues, some of which are of great relevance to the teaching and learning of mathematics. (For a good summary of the literature from the point of view of mathematics education, see the paper by Clements, 1981-82). Some of the writing is much taken up with problems of definition, but this kind of fundamental philosophic approach may not be the most important one for educators. More recently the emphasis has been on the practical issue of usefulness, and the contribution mental imagery can make to learning and to mental functioning generally. For educators an approach of this sort is likely to be more relevant, since the crucial questions for them are about the effect of generating and using pictorial representation on such issues as learning, problem-solving and understanding.

For example, the interrelationships among such phenomena as memory, understanding and mental imagery are complex and there is a danger that forms of dependence and priority will be assumed. For example it might be argued that mental imagery relates closely to memory and to communication, and that these in turn relate to understanding. But such relationships are at best probable since we know, for example, that things can be remembered which are not understood and, to paraphrase Shelley, pictures can communicate before they are understood.

In 1971 *Mental Imagery and the Child* by Piaget and Inhelder was published in English. This produces some useful distinctions which place visual or pictorial imagery in the context of other forms of imagery. They argue that all communication uses 'imagery' in the form of a variety of symbolic representations such as words (spoken or written), movements and gestures, mathematical symbols and pictorial forms. The relation between the human thought-processes themselves, and the images used to present or represent them, is usually conventional, and the form of representation is either arbitrary or a matter of preference. The result is some form of system 'whereby each concept, which would otherwise remain abstract, is coupled with a figuration that gives it "examplarity"' (p11).

145

The particular case of pictorial mental imagery, as opposed to, for example, aural, or symbolic, involves problems about how pictorial images are input, processed and stored in memory, and about the reverse processes of accessing these images. These questions are related to more general questions about how knowledge is represented in the mind and this topic is currently of great interest in the field of artificial intelligence, which itself has clear links with the use of computers in education — thus sort of completing the circle. For example, Self (1985) writes: 'Computers process representations of information — which is not the same thing as information, still less is it knowledge, and even less education' (p57).

It also seems to be the case that much of the formal writing and research about mental imagery, especially in psychology, concerns itself with relatively simple or readily available examples arising out of immediate and unsophisticated sensory stimulation. When reading descriptions of this research there is an immediate contrast between this simple 'experimental' imagery and the dynamic and complex and multi-faceted quality of much of the child's experience of images in the world generally, and particularly the modern world of colour and graphics derived from such things as video games, TV, and so on. Shelley Turkle (1985) describes the imaginative personal worlds of fantasy which many children create alone or together through the new electronic media. So it may be the case that the material examined by researchers in the field of imagery is a poor-quality, artificially faded, and dim form of reality. This may explain the poverty of some of their findings. When we turn to the more complex and sophisticated pictorial imagery which modern electronics makes available to children, it becomes likely that this is a way of representing or symbolizing ideas that research has not yet begun to take account of.

Mental imagery and mathematics

One of the few truths about mathematics education that we must continue to take account of is how little we know about even the most elementary problems within it. Alan Bell is making the same point with the phrase 'most of what is taught is not learned', Bell, 1981, p28. This situation probably arises in part from the nature of the subject, but it also follows from a lack of knowledge in more general terms about learning and mental processes: 'The central problem of math education is, in my view, our almost total ignorance about the cognitive processes involved in the acquisition of mathematics . . . We do not know the relationship between the modes of instruction and the internal representation of mathematical objects and concepts' (Nesher, 1981, p27).

It is not surprising that, in the absence of such knowledge, teachers use their intuition and judgements and experience when presenting mathematical ideas. For many this appears to involve a lot of pictorial forms. So textbooks, workcards, project books and all the other formal forms of classroom presentation of mathematical ideas use diagrams and pictures and figures as ways of representing and of presenting ideas from all parts of mathematics. The ubiquitous circles for fractions, the numberlines, Venn diagrams, tessellations, coordinate systems, and so on, are all used and thought of as ways of easing children into complex abstract ideas. They are clearly thought of as being helpful, as bridges between the words and the symbols of communication, and the ideas themselves. Partly this is because of the difficulty of communicating mathematics: its symbolic language is synoptic and conventional and usually arises from specific historical contexts; ordinary language alone is not always adequate since it too is conventional and abstract and often personal. So other forms of communication are sought and used.

This is particularly true of geometry, at least in the early stages, since the development of geometrical intuition seems in a natural sense to be related to and helped by spatial and pictorial imagery. However, much of the geometry currently taught in schools takes a literal or algebraic form rather than a pictorial one, and so the symbolic imagery of pictures as a way of communicating ideas and concepts becomes increasingly less useful and less direct. To some extent this has been caused by the difficulty of finding forms of pictorial imagery to represent some of the more abstract new 'geometrical' ideas, such as invariance and groups of transformations. This is true also of much modern physics and Piaget and Inhelder (1971) refer to the general problem as 'a crisis of imaginal representation' (p13). But it may also be the case that the kind of geometry, and indeed the kind of mathematics, that schools will turn to in the future will show the influence of the microcomputer in that pictorial imagery will be an important constituent.

Papert in his book *Mindstorms* (1980) describes the basic problem as one where the normal everyday culture of children has no paths into or links with the essentially formal ideas of mathematics (or science generally). Such subjects have intellectual pre-suppositions for which children have little cultural preparation, and many of these are implicit or unrecognized, and are counter-intuitive. He cites the Piagetian child-world of pre-conservation as an example. For Papert the microworlds of computational geometry provide children with 'objects to think with', that is they represent ways of making links between the theories of the child and the world of mathematics. For example, he writes at length about the process of going from the child's physical walking of a circle to the LOGO procedure for making the turtle draw a circle. From our point of view it is interesting that Papert has no doubts about the importance of the powerful visual imagery which his system exploits.

An experiment with the computer

There is a number of ways of using the computer to present children with visual images of mathematical ideas. One way is to avail of the power of the machine to present what Papert calls 'powerful ideas'. This means that mathematical concepts normally thought advanced can be introduced as pictorial ideas and dealt with on an intuitive rather than on a formal level.

In order to investigate some aspects of this issue an experiment was carried out with a number of groups of schoolchildren of various ages, using a set of three programs, specially written by the author, which presented the generation of the sine curve in the form of an animated 'film'. The main purpose of the programs was to demonstrate the notion of a circular or periodic function using colour, graphics and movement. This involved showing the programs to children and discussing with them what they might mean. The groups consisted of five or six children and an observer, and usually met for at least an hour and a half.

The programs did not try to teach about sine curves in any direct way. No attempt was made, either in the programs or by the observer, to explain things or to use descriptive or technical words. The picture was the only message, and this simply showed how a radius revolving round a circle could be used to generate a full cycle of a sine curve. There was no physical interaction in the sense that the watchers could not change what was being presented, except for two things: they could freeze the action at any time to examine a particular 'snapshot' and they could run the 'film' over and over again. The programs are not written in LOGO, but they follow Papert in that they are meant to provide children with a bounded but intellectually rich 'microworld', which uses pictorial imagery to develop intuitive understanding of otherwise difficult and abstract ideas. They therefore arise directly out of the Piagetian theory that children build their own intellectual structures, as a result of their constant interaction with the environment.

The first program presents a circle on the left of the screen with a slowly revolving radius which is the hypotenuse of a right triangle. As the radius revolves the other two sides of the triangle change size. At the same time ordinates representing the length of the vertical side of the right-angled triangle appear on a graph alongside the circle, thus producing a sine curve. (Figure 27 shows a snapshop from this). Nothing, apart from these lines and curves, appears on the screen except the words. 'To stop or start the movement press the SPACEBAR'. When the full curve has been drawn the words change to, 'Do you wish to see that again?', and so the user can choose to look at this film again, or to go on to the next one.

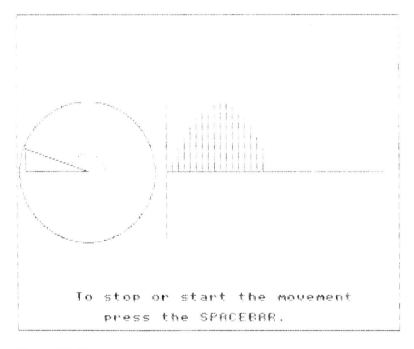

Figure 27. First program's generation of a sine curve

The second program again shows the generation of a sine curve, exactly as before but more slowly. As well a box is added at the top of the screen containing information about the size of the angle traced out by the radius, and a statement about the value of the 'sine' of the angle, like this, (see also Figure 28):

Angle in degs is   50
Sine of angle is   0.766

149

No attempt is made to explain with words what any of this means. As the radius rotates the angle changes (in increments of 10) from 0 to 90 to 180 and so on; at the same time the 'sine' changes from 0 to 1 to 0, and so on. Again it is possible to watch this process unfolding as often as is desired.

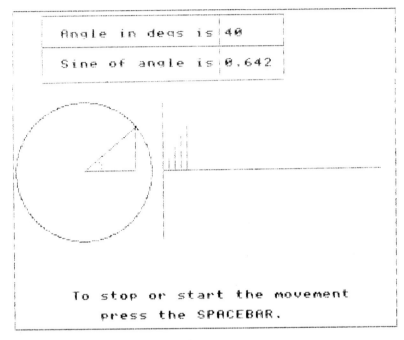

Figure 28. Second program's generation of a sine curve

The third program again draws the circle and a corresponding complete cycle of the sine curve. This is done quickly as a sort of summary of the previous processes. Four points on the curve are then indicated at each of which the numerical value of the sine of the angle is the same. The first angle is chosen at random between 0 and 90 degrees. An example is shown in Figure 29, and the angles can be thought of as x, 180 − x, 180 + x and 360 − x. The value of x is shown and the user is then asked to input the values of the other three angles.

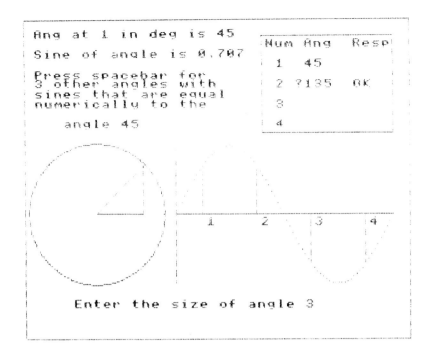

Figure 29. Third program showing problem posed for finding values of angles

These programs have been used with a large number of groups of children between the ages of 9 and 13. Although this work was thought of as an experiment, this was a pilot stage, so no attempt was made to make any measurements of children's responses or to collect any formal empirical research data. It was felt necessary at this stage to gather impressions and to use these, both to improve the programs and to act as a guide for future research planning. Without any attempt at differentiating specific groups, some responses to these programs are now described. These are deliberately kept short and undetailed and are meant to give a flavour of the experience. Once the program was running the observer did not interfere in the decision-making about the way forward. The children at all stages decided for themselves what to do next.

Perhaps the most surprising thing to emerge initially was how often children, irrespective of age, wanted to watch the 'film' run. Again and again they chose to watch it through, to begin with almost

151

in silence. Eventually this was replaced with commentary and discussion which had two specific dimensions. The first was randomly descriptive and referred to things which they imagined the pictures reminded them of; the second involved making definitive statements about aspects of the curve, which seemed none the less to have the status of hypotheses. 'If that bottom half were moved over to the left it would be a circle.'

Their desire to see the process repeatedly seemed like the child Edith Biggs once referred to who repeatedly checked, over and over again, that three 'foot rulers' fitted exactly onto a 'yard ruler'. Certainly it would be unlikely that a teacher, introducing these ideas in a conventional way, would show them as many times as children seemed to need or want. Perhaps one of the reasons why children remember and understand with such apparent ease so many things from their own culture, is that they look at them or listen to them with a frequency which teachers would think unnecessary, or wasteful of valuable time, and which they might fear children would be unlikely to accept. But if it is realized that we know little about how children learn, the computer seems to provide an opportunity simply to watch and listen to the process.

In all of the sessions there were quite strong levels of concentration and interest. No one complained about the ideas being hard or unpleasant, and there was obviously no perception of the lesson being about mathematics. Mostly to begin with they concentrated on the completed sine curve on the right of the screen and discussed its shape. They debated as to whether the two halves of the curve would together make a circle and usually agreed that they would not. They used words like 'oval' and 'egg-shaped' along with much desultory talk about symmetry and angles.

They then usually turned their attention to the changing triangles on the left. These confused them a great deal to begin with, and they did not seem able to isolate particular lines and watch how they behaved. It was only after watching the film many times that the vertical projection or 'sine' was perceived as changing in size as it moved round. For a long time it was as though it was seen each time as simply a different line on a different triangle. The revolving radius was also perceived as a sequence of lines in different positions,.until someone used the image of the big hand of the clock: and it was only then that the word radius was used. However once the lines were seen as moving and changing in size, the connection with the curve began to emerge. (In most cases they were surprisingly reluctant to say that a line became of zero length. It became smaller, and then it became bigger.) But it was usually some time after this that the notion of a changing angle at the circle centre took their attention. Once it did, the normal reaction was to talk about the relationship between the changing angle, the changing lengths, and the evolving curve. This then led to an apparently strong intuitive notion that the curve should also be a circle, and there was considerable discussion about why it was not when it clearly ought to be. This notion was common and persisted and no explanation was offered which satisfied everyone.

When pupils moved on to the second program the appearance of the numbers did not, for a long time, make any sense to them. (Requests to be told what the word 'Sine' meant were not responded to.) This film was also looked at a large number of times, if not quite so often as the first one. As before, the level of sophistication in the analysis of what was happening increased dramatically over a comparatively short period until eventually they were giving a substantially correct analysis of what was happening.

The movement towards this correct analysis involved what can only be described as hypothesis-making. To begin with this took the form of often quite lengthy explanations for particular developments. These explanations were as often as not incomplete or inaccurate and were almost always seen to be so. To begin with they were not much connected with those that preceded them, but as consensus developed about particular aspects, the discussion focused down quite quickly on a correct analysis.

The third program represents a 'test' related to the first two programs. In this it is hoped that the significance of the numbers 180 and 360 will be seen, and will be exploited to find the three related angle sizes. The success rate here was almost total although it is not intended to make too much of this since a number of alternative explanations are available. Children were then asked to draw curves like this, to choose an angle for themselves and to calculate the remaining three angles as before. Again all made some attempt at this and most performed calculations that were accurate in intention.

Clearly this experiment represents one small example only and cannot be used as a base for generalizations. It does seem to suggest that mathematical ideas often thought of as quite difficult can be made available to children through the medium of computer-generated graphics. In this it reflects much of what Papert writes about the potential power of children's thinking, and about the need to provide them with intuitive links between their own world and new ideas. It also suggests that some aspects of the spirit of LOGO can be transferred into other contexts using other tools. Perhaps the most interesting result is the support given for the idea that the learning procedures adopted by children may be different from those that a teacher might expect. For example, the need to repeat a visual experience, or indeed any experience, a large number of times is not unexpected in itself. But repetition in the school context is usually associated with rote-learning and the failure of understanding. In this case, however, the repetition seemed to have two functions. The first was to allow the constant testing of hypotheses generated within the groups. The second allowed the overall concept to be built up out of a sequence of small individual understandings. The logical relationships between the various parts of the 'film', and the diagrams within it, were seen and taken account of by different children and groups of children in different ways and in different orders. In this situation the idea of 'individual differences' takes on a new significance in that it relates to the singular and unique way any human puts together the inputs from the world to make a private understanding.

The social dimension of the experience also seemed to be important. Using the computer with others led to constant negotiation about meaning and relevance. Much of the talk was irrelevant, or 'wrong' in the sense that it led to false trails. But it seemed that one of the most powerful ways to find a correct trail was to try a number of false ones, and to test these against the experience, in this case the film.

The importance of the pictures, or the mental imagery, is impossible to estimate in any scientific way, but it would be difficult to imagine any other way of getting nine-year-old children to talk sensibly about circular functions. Learning from pictures, or learning to read pictures, may be a different form of cognition, one which exploits intuitive notions about the world, and which reflects the child's ordinary world culture of television, videos, video games, and so on. Education may have to come to terms with this in the future, in relation to most of its aspirations and procedures, including learning, teaching and the curriculum.

References

BELL, ALAN, 'A research program for mathematics education', *For the Learning of Mathematics*, 2, 1, 1981

CLEMENTS, K, 'Visual imagery and school mathematics', *For the Learning of Mathematics*, pp2-9, November 1981, and pp33-39, March 1982

GATTEGNO, C, 'Mathematics and imagery', *Mathematics Teaching*, 23, pp22-24, 1965

NESHER, P, 'A research programme for mathematics education', *For the Learning of Mathematics,* 2, 1, 1981

PAPERT, SEYMOUR, *Mindstorms: children, computers and powerful ideas*, Harvester Press, Brighton, 1980

PIAGET, J and INHELDER, B, *Mental Imagery and the Child*, Routledge and Kegan Paul, 1971

SELF, JOHN, *Microcomputers in Education*, Harvester Press, Brighton, 1985

TURKLE, SHELLEY, *The Second Self: computers and the human spirit*, Granada, London, 1984

Investigating graphs and the calculus in the sixth form

Norman Blackett, Kenilworth School and David Tall, University of Warwick

The microcomputer presents unrivalled opportunities to help students understand mathematical concepts with its fast numerical processes, moving graphics and interactive facilities. However, the reality of the mathematical classroom is that too few micros are currently available to allow students adequate access. The present article describes the use of a single BBC computer in a lower-sixth classroom for drawing graphs and providing insight into the calculus. The micro was available for most of the mathematics periods during the year and used whenever it seemed appropriate.

A large part of an A-level pure mathematics course consists of:

(i) an introduction to real functions and investigations into their behaviour, usually incorporating a pictorial approach
(ii) practising algebraic techniques and manipulation of trigonometric identities
(iii) introductory calculus with particular emphasis on derived functions and antiderivatives of combinations of polynomials, trigonometric functions, exponentials and logarithms
(iv) elementary ideas of proof.

We were concerned to see how a graphical approach could contribute to effective teaching in each of these areas. Much of the initial work appeals to pictures of functions and pictorial illustrations of finding the derivative. But because of the limitations of blackboard and chalk and static pictures in textbooks this quickly gives way to algebraic processes. Our idea was to use the computer to give an understanding of the geometric ideas. For this purpose we developed and tested the graph-drawing program 'Supergraph' and the suite of programs 'Graphic Calculus'. The plan was to provide facilities for graph-drawing that were so flexible that the computer could be switched on whenever we felt the urge to ask a 'what if' question in terms of drawing a graph.

The use of the computer proved to have a profound effect on the relationship with the sixth-formers. They were much more willing to discuss ideas illustrated on the computer as they typed in expressions themselves, than they might have been if the concepts were introduced with the authority of a teacher's 'talk and chalk'.

They were able to conjecture what might happen, suggest possible formulae for derivatives, test them with the computer and investigate ideas experimentally before they were proved formally. Their insight into the geometrical ideas proved far greater than comparable groups of students who had not used the computer, without their ability to do the algebraic manipulations being impaired. In a word they did mathematics *actively* rather than simply learned passively at the teacher's instigation. They learned not only mathematics; they learned how to learn.

Elementary graph-sketching

Some syllabuses postpone graph-sketching until after differentiation to take advantage of the derivative in determining maxima and minima. As our approach to differentiation depended on an experience of graphs, we studied polynomials and trigonometric functions in four stages:

(i) sketching graphs of polynomials and rational functions
(ii) derivatives of polynomials and powers (including negative and fractional powers)
(iii) trigonometric functions in radians and trigonometric relationships
(iv) derivatives of trigonometric functions.

'Supergraph' is such a flexible computer program that it can be used by any level of pupil or teacher. One version of the program allows superimposition of any number of cartesian, parametric or polar graphs and another sacrifices the parametric and polar options for a wide variety of other facilities, including tangents, normals, lines, zoom options, etc. Both versions allow normal algebraic input (with cursor movement for inserting powers) and letters other than x,y are taken as constants. For example a straight line could be typed as

$y = mx + c$

or a quadratic as

$y = ax^2 + bx + c$

and the constants may be varied to superimpose variations on the graph including families and envelopes. For instance, Figure 30 began as the line

$y = mx + c \, (m = 1, c = 0)$

With a succession of graphs fixing m and increasing c by 1 each time, we see that the graphs are all parallel. Alternatively, fixing c and varying m shows that the graph always passes through the point given by x = 0, y = c. By making m negative it is possible to see a straight line graph with negative gradient, or m = 0 gives a horizontal graph.

y=mx+c:c=5,m=1.

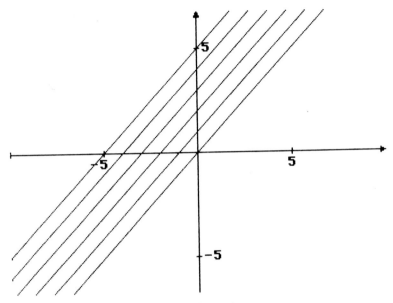

Figure 30. Succession of graphs drawn from y = mx + c

We did not need to have any special book of directions to use 'Supergraph' in graph-sketching. We simply worked from our usual (SMP) textbook and illustrated graphs as they turned up. One nice discussion we had concerned the graph of

y = 1/(x − a) (x − b)

(note that 'Supergraph' gives division precedence over implied multiplication. We first drew the graph

y = (x − a) (x − b)

for a = 1, b = 2 and then superimposed the graph of the reciprocal (Figure 31). We could see that the two graphs had the same sign but where (x − a) (x − b) was zero the reciprocal had asymptotes.

158

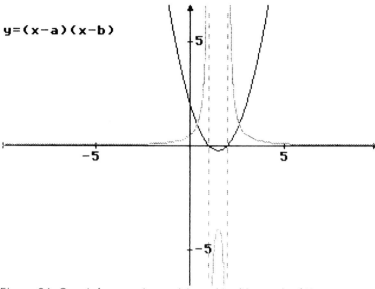

Figure 31. Graph for $y = (x - a)(x - b)$ with graph of the
reciprocal superimposed

By considering the sign of the product of the terms for $x < a$ and
$a < x < b$ we could see that the algebraic expression $1/(x - a)$
$(x - b)$ is very large and positive before a and very large and
negative after it. We reinforced this message by substituting
$x = a + h$ to get the expression

$$y = 1/h(a - b + h)$$

and considering the sign for small values of h. In this way we linked
the algebraic ideas with a pictorial representation.

The interesting investigation was to take $a = b$ and see what
happened. Before drawing the graph, the algebraic substitution
$x = a + h$ gave

$$y = 1/h^2$$

and it could easily be seen that the graph should be large and
positive both before and after a. This was confirmed by drawing the
picture. From here it was a simple matter to conjecture what would
happen to

$$y = 1/(a - x)^3, y = 1/(a - x)^4 \ldots$$

159

and test the result on the computer. Likewise we could look at a combination such as

$$y = 1/(a - x)(b - x)^2.$$

In some cases the students were not sure of the scale of the picture, so they used automatic scaling to get a sighting before drawing the graph using equal scales. This helped them to think about the kind of ranges it is appropriate to draw the graphs. They soon started checking the graph sketches given in the textbooks and were amazed to see how inaccurate they were. It threw the whole question of graph-sketching by old methods somewhat into disrepute!

The gradient of a graph through magnification

Not all students work at the same pace and one day Allan finished his exercises well before the others. He was given the program 'Magnify' from the Graphic Calculus Pack and told to draw some graphs, magnify them, and report what happened. Two or three minutes later he said 'They look less curved'. He was joined by other students who tested their conclusion by magnifying other graphs. They were asked to explain their ideas to the rest of the class.

Was this property of graphs always true, or did it sometimes fail? Discussion followed. Most of the class were convinced that it was always true. But was it?

They tried all sorts of expressions to cause the idea to break down and failed. The graph $y = abs(x)$ (the 'absolute value' or 'modulus' of x) was suggested by the teacher. It clearly had a 'corner' at $x = 0$. Other graphs, such as $y = abs(x^2 - 1)$ also had corners.

Thus it was that the students began to appreciate that some graphs looked straight under high magnification and some did not. Another program in the pack was used to show the 'blancmange function', which is so wrinkled, no matter how highly it is magnified, it never looks straight. An amusing and fascinating discussion followed. Clearly there were many curves in nature that failed to 'look straight' when highly magnified. Even a ruler is wrinkled viewed under a microscope.

The next lesson we looked at the gradient of a graph through this approach. If the graph had the special property of approximating to a straight line under a magnifying glass, we could talk about its gradient in a small segment as being the gradient of the magnified (almost) straight portion.

This proved easy to see (literally!) with the computer. Although we had a routine to display the limit of the chord from a to b as b tended to a, we found it much more profitable to attack the gradient of the graph as a dynamic picture. The idea is simple: draw the chord from x to x + c (where c is small) and then let it click along the curve as x increases, and at each click, plot the gradient of the chord as a point, leaving a trace of gradient points behind (Figure 32).

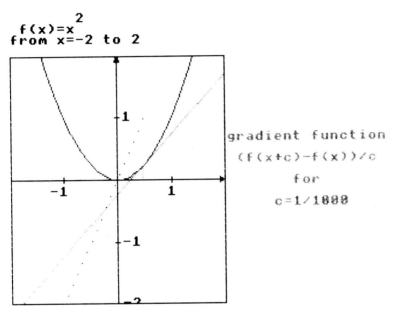

$f(x)=x^2$
from x=-2 to 2

gradient function
$(f(x+c)-f(x))/c$
for
$c=1/1000$

Figure 32. Gradient of a chord plotted as a point

The gradient of $y = x^2$ proved to be a revelation; it was virtually a straight line when c was small!

By checking the algebra, we found the gradient from (x,x^2) to $(x + c, (x + c)^2)$ to be

$$\frac{(x + c)^2 - x^2}{c}$$

$$= 2x + c \text{ (for } c \neq 0)$$

and it was evident that for small c the gradient would approximate to $y = 2x$.

The software included a routine to type in $y = 2x$ and make a comparison with the gradient. Give or take a pixel, the graphs proved to be identical.

Interestingly enough, the odd pixel or so difference provoked an interesting discussion in which it was realized that the graph-drawing routines only calculate a few points and join them up by lines. As the picture is actually made up of dots of light (less than 200 by 200 in the graph-square) it is self-evident that there are likely to be inaccuracies in drawing anyway.

Putting on a show

That evening was the Open Evening, giving parents the opportunity to see the kind of work going on in the sixth-form centre. As usual the science departments had all their experiments on display and mathematics was keen to compete. The obvious weapon was the computer. On the spur of the moment we organized small groups of students from the calculus set to come and investigate the gradients of graphs in front of the visitors. Each group was given a sheet of challenges. They knew the gradient of x^2 was $2x$, could they guess the gradient of x^3, and when they had done that, could they guess the formula for x^n? This was easy. The first group drew the gradient function (Figure 33) and saw that it was a U-shape but clearly not x^2, so they guessed from their earlier experience that it might be x^4. By superimposing this graph and comparing with the gradient they saw the error of their ways and immediately moved on to try $2x^2$, then refined their guess to $3x^2$. It worked!

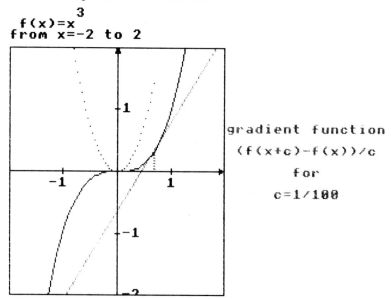

$f(x)=x^3$
from x=-2 to 2

gradient function

$(f(x+c)-f(x))/c$

for

$c=1/100$

Figure 33. Gradient functions drawn from guesses for x^n

A quick conference and they guessed that the gradient of x^n must be nx^{n-1}. Any mathematician might think they would try out this expression systematically with n = 3. Not a bit of it: their first check was n = 33. The computer creaked a bit and dutifully drew the graphs and confirmed their suspicions.

The next task was to try the formula for n = −1, −2 and others such as n = −1/2 and it went according to plan.

Following this the challenges leapt ahead with a brief description of angles in radians and a challenge to find the gradient function for sinx, drawing the graph from −2π to 2π. (The programs allow π to be typed in as pi which translates to Greek before the operator's very eyes.) They cracked the problem immediately on seeing the gradient drawn and guessed the gradient of cosx too. They were pressing on eagerly now.

The next bastion to fall was the discovery that 2^x and 3^x had similar shaped gradient functions and that somewhere in between was a number k such that the derivative of k^x is again k^x. Then they conjectured the derivative of the natural logarithm of x, and even managed that of ln(abs(x)) . . .

This was expected to take them some time, but it was all over in a few minutes with them asking for more. They were challenged to find the gradient function of tanx (Figure 34). The response was that the gradient of tanx looked like the square of the tanx graph, except it was 1 where $(tanx)^2$ was 0, so the gradient was probably $1 + tan^2x$. (They had to be shown that the program required the input as $1 + (tanx)^2$. It is not possible to fit every refinement into the tiny BBC memory.)

f(x)=tanx

from x=−3π/2 to 3π/2

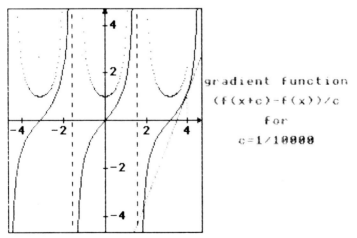

Figure 34. Gradient functions drawn from guesses for tanx

163

They followed this up with other curves to their interest and eventually admitted defeat when they could not guess the formula for the derivative of the semicircle $(1 - x^2)^{1/2}$.

Other groups had similar successes. The difficulty was keeping the earlier experimenters away to give the later ones a chance.

New methods of approach
We followed up with more student use of the programs in the regular mathematics classes. We quickly realized that with one micro for 16 students we needed a more systematic approach for its use than at the Open Evening. An extra computer at this stage would have been invaluable. With only one computer we divided the class into groups of three or so. As they went through a set of regular differentiation exercises, they took it in turns to go to the computer, type in their functions and check that their result actually worked. The rules were simple: they were assigned exercise numbers, the first group to the first question, the second to the second, and so on, until all the groups were exhausted, then the first group had the next question and the process repeated. They could go up at any time when they had completed a given exercise and the system of staggering the questions meant that they could get on with the exercises until there was a free space to try the computer. They were told to bring any interesting features of the graph to the attention of the whole group, so occasionally work stopped to see what they had found.

They soon got used to the idea of visualizing the gradient function as a dynamic process, looking along the graph, seeing the gradient change, and visualizing the gradient as another graph.

Fitting in with the Physics Department
One day several of the students who had missed the Open Evening experience came to the class saying that the Physics Department had started on 'simple harmonic motion' and they did not understand the derivatives of sine and cosine. We switched on the computer and went through the exercise for them. Even though we had yet to cover the trigonometric functions in radians, this proved most helpful and they professed satisfactory insight into what was going on. Except Brian: he did not see what all the fuss was about. What he wanted was to be told the formula so that he could learn it and pass the exam. All this computer stuff was a waste of time.

More graph-sketching
We returned to graph-sketching to study the trigonometric functions. The students were encouraged to sketch the graphs of sine and cosine by the time-honoured method of drawing a circle radius 1 and

transferring the values of y = sinx to an x − y graph as the angle x turned (several times) through full circle (Figure 35). We could have programmed a computer to do the drawing for us, but there is something essential in the act of reading the values round the circle and physically transferring them to the graph. In the same lesson they drew cosx, then moved on to the functions sin2x, 2sinx, and other similar expressions.

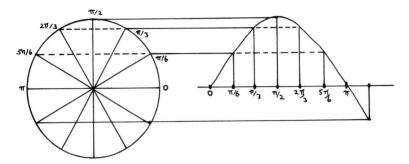

Figure 35. Sketch of the graph of sine

Once again the physical act of drawing proved indispensable. For instance, y = 2sinx requires the y − value for y = sinx to be read, multiplied by 2 and the new graph plotted whilst y = sin2x requires the y-value to be read from y = sinx, then the x-value halved to give the new graph. None of this action would have come out on the computer.

But when the initial graphs had been drawn and the physical effort made, it was a luxury to sit back and draw sinx, 2sinx, 3sinx, to see the growing y-values, then sinx, sin2x, sin3x to see the reducing x-size. After studying the angle formulae such as

sin(a + b) = sin(a)cos(b) + cos(a)sin(b)

which we 'proved' by the 'modern' matrix method, we found that it was met with some resistance. But superimposing the graphs of

sin(a + x)

and

sin(a)cosx + cos(a)sinx

at least gave some feeling of confidence. We had an interesting experience with drawing

y = asinx + bcosx.

165

What would the graph look like, say for a $= 3$, b $= 4$? The students had no idea. Even the physical sketching proved hard. So we compared this with the formula for

ksin(p + x) = kcos(p)sinx + ksin(p)cosx.

We required

a $=$ kcos(p), b $=$ ksin(p)

which gave

k $= \sqrt{(a^2 + b^2)}$, tan(p) $=$ b/a.

So we drew y $=$ asinx + bcosx (for a $= 3$, b $= 4$) and superimposed y $=$ ksin(p + x) where

k $=$ sqr($a^2 + b^2$), p $=$ atn(b/a).

Lo and behold, the superimposed graph was, pixel for pixel, identical with the original.

Local behaviour

The students had grave difficulties sketching the graph of

y $= \sin^2x$.

Imagining the graph of y $=$ sinx and squaring, most of them ended up with a sketch with cusps where the graph met the x axis (Figure 36). When the true graph was drawn, they expressed some mild surprise. But it was easy to focus their attention on the fact that the graph of y $=$ sinx near the origin was much the same as y $=$ x, so the graph of y $= (\sin x)^2$ must be similar to y $= x^2$. Hence the rounded shape at the origin, and also at every multiple of π.

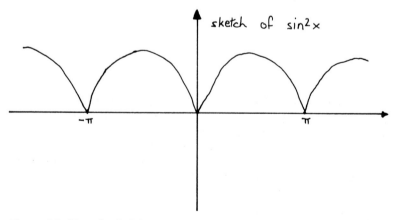

Figure 36. Sketch of sin²x

166

Following up this idea, what would the graph of $y = \sin^3x$ look like? They sketched it and compared it with the computer picture which proved to look much like x^3 at every multiple of π (Figure 37). A picture in a book only conveys part of the feeling of the graph growing in front of your eyes on a computer screen, especially when that graph is drawn to your bidding.

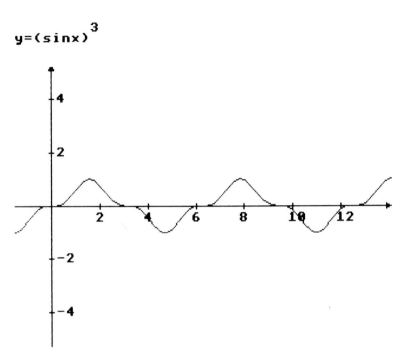

Figure 37. Computer picture of graph of $y = \sin^3x$

Proof of the formulae

When we came to derive the formulae for the derivatives of sine and cosine we were in the fortunate position of knowing the answers before we went through the trigonometric manipulation. The students' experience of the gradient program stood them in good stead and they could see that the gradient from x to x + h would be much the same as the gradient from x − h to x + h. So the approximate gradient was

$$\frac{\sin(x + h) - \sin(x - h)}{2h}$$

167

The fabled formulae transformed this to

$$\frac{\text{sinxcosh} + \text{cosxsinh} - (\text{sinxcosh} - \text{cosxsinh})}{2h}$$

or

$$\text{cosx} \frac{\text{sinh}}{h}$$

Their experience with the local behaviour of the graph of sinx showed them that, near the origin,

$$\frac{\text{sinx}}{x}$$

was approximately 1 and this was reinforced by further discussions of well known ideas. For small h therefore the gradient of the sine curve approximated to cosx, which was as expected. The cosine curve followed similarly, including the magic minus sign in the derivative − sinx. But this minus sign now had a physical interpretation: when the gradient function for cosx was drawn it simply turned out to be the graph of sinx upside down (Figure 38).

f(x)=cosx

from x=-8 to 8

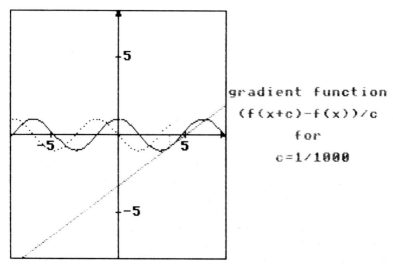

Figure 38. Computer picture of gradient function for cosx

Gradient sketching

By this time, without being taught explicitly the students were very good at sketching gradients. Those following a standard A-level course faced with the graph in Figure 39 might consider drawing its gradient in a number of stages.

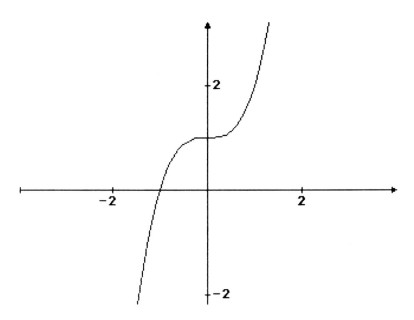

Figure 39. Computer picture of gradient from which derivative must be sketched

'The graph looks like $x^3 + 1$, so its derivative will be $3x^2$ and a sketch of this can now be drawn.' But a student with a dynamic view of the gradient would say 'The gradient starts off large and positive, it diminishes rapidly until the gradient is zero at the origin, then increases just as rapidly again beyond this point.' Thus a sketch of the derivative could be drawn in a single procedure. Slightly more complicated graphs which could confuse many students (because they could not guess the formula) would be equally well handled by those students with a dynamic image of the gradient. In tests between control and experimental groups the improvement was significant.

Handwaving

Using the programs for differentiation had an unforeseen effect. They outgrew their usefulness very quickly. This is not to say that they were not used on occasions in the later stages because they were always there to draw a difficult derivative. But the students could now interpret a static picture on the blackboard or a vague handwave in the air as a dynamic representation of a gradient function. When we came to discuss the properties of maxima and minima they responded immediately to questions about the gradient of the graph at, before or after the turning points.

Antidifferentiation and the arbitrary constant

The idea of antiderivative is usually blessed with the name 'indefinite integral' in our A-level. It is a misnomer if ever there were one. It means that we know f(x) and we want to find a function A(x) such that A'(x) = f(x). We can look at this two ways: the standard way (I want to look for a formula which differentiates to give f(x)) and the pictorial way (I know the gradient f(x) of the graph y = A(x), can I draw a suitable graph for A(x)?). It does not take long to consider the pictorial approach and it has an unforeseen benefit.

Through an array of points in the plane we draw short line segments of gradient f(x). Then y = A(x) is obtained by following along the directions given (Figure 40). Because the gradient depends only on x, not on y, the possible graphs through a vertical line all have the same direction. Thus the solutions differ by a constant.

f(x)=1/x

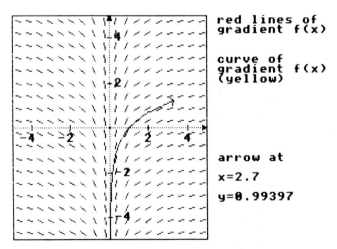

Figure 40. Computer picture of short line segments of gradient f(x)

However, if the function f(x) is undefined somewhere, for instance, f(x) = 1/x is undefined for x = 0, then we can not trace the solutions through this value of x. A solution of A'(x) = 1/x by tracing the directions will lie on one side of the origin only and not cross it. Thus the 'added constant' only applies on one side of the origin. It is perfectly possible to shift the parts of the solution on either side of the origin up or down by different constants. Thus we get the truth about the 'arbitrary constant': it is only valid in a *connected* part of the domain of f(x). Too difficult for students to understand? Try it and see!

Effects on algebra

The algebraic techniques necessary to answer standard sixth-form questions were developed in conjunction with the appreciation of visual concepts. Interestingly, it became obvious that skills in one of these areas did not necessarily imply skill in the other. Colin, who could soon talk fluently and intelligently about the behaviour of functions near asymptotes, sketch derived functions, appreciate differentiability and the geometry of translation and reflection of graphs, found great difficulty in successfully developing three or more lines of algebra. By contrast David, who found little difficulty with algebraic manipulation, often being amongst the first in finishing an exercise, needed several prompts to appreciate the picture of some of the functions he was dealing with. However, the results shown graphically certainly provided motivation for proceeding to a more formal algebraic approach, with the sequence

illustration $\rightarrow$ conjecture $\rightarrow$ proof

being the most successful method of interesting the class in the nature of algebraic proof such as the derivative of x^n.

The computer graphics also provided a good way of checking results. The sense of triumph when $y = (1 - x^2)^{1/2}$ was differentiated by the chain rule, the result drawn on the computer and found to be identical to the superimposed computer-drawn derivative far exceeded that usually experienced by looking up the answer in the answer book!

In a similar way, innocuous-looking algebraic errors in using the formulae for differentiation were shown to produce markedly different graphs from the actual derivative, thus underlining the importance of careful algebra.

Algebraic difficulties are such a recurring problem that they need a total rethink in how they are introduced earlier in the curriculum. (See 'Playing algebra with the computer' on p59.)

Insight into area

The next stage in the course was to investigate the area under a graph by summing rectangles. The computer picture of the process quickly showed the effect of increasing the number of rectangles by reducing the width of each rectangle to gain a better approximation to the area under the curve. The program shades the rectangles differently according to the sign of the area (Figure 41). Before any calculation was performed by the class they were able to appreciate how the signs of the ordinate and the step combine to give the sign for the area of each rectangle.

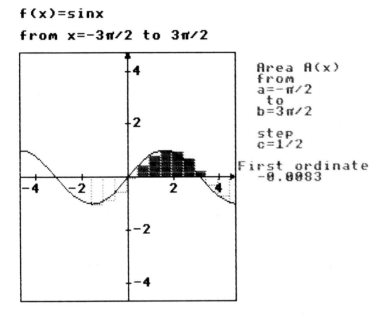

Figure 41. Computer picture of summing area under graph by drawing rectangles

When it came to performing the algebra necessary to sum the area of rectangles under $y = x^2$, there was the usual trouble with Σn^2, but they did make more sense of the limit of the sum as the width of the strip tended to zero after seeing the calculation performed on the computer. After looking at the results for the area under $y = x$ and $y = x^2$ from $x = 0$ to $x = b$ they were willing to conjecture the

172

result for the area under $y = x^3$ for $x = 0$ to $x = b$. The algebra required to show this result would have certainly lost many of the group, but the computer was able to support their conjecture.

At this point we looked briefly at the other numerical methods for estimating area given in the package, such as the trapezium rule, Simpson's rule and the midordinate rule, to compare results. The class were soon able to see the relative efficiency of these methods and make good guesses as to whether they would overestimate or underestimate the area under a given curve.

We progressed to the standard notation for integration. After seeing the rectangles being drawn, it was a relatively simple matter for the class to appreciate that the area from $x = a$ to $x = c$ is the sum of the areas from $x = a$ to $x = b$ and $x = b$ to $x = c$ and, more importantly, that the area from $x = b$ to $x = a$ is *minus* the area from $x = a$ to $x = b$.

They had conjectured the result of the Fundamental Theorem of Calculus for some easy polynomials, had it confirmed for more difficult examples, and were now ready for some form of proof of the general result.

The idea behind the fundamental theorem was illustrated on the computer by stretching the x-range of a graph over a small interval. The y-range was left unchanged and the increase in area from x to $x + h$ was examined. We chose $f(x) = \sin x$ as a typical illustration, using ranges from $x = 0.99$ to 1.01 and $y = -2$ to 2 (Figure 42). It can be seen that a horizontal stretch flattens out the graph to give the area calculation in the form

$$A(x + h) - A(x) \simeq hf(x).$$

The more the x-range is stretched, the flatter the graph gets and the better the approximation becomes. By rewriting it in the form

$$\frac{A(x + h) - A(x)}{h} \simeq f(x)$$

and allowing h to tend to zero gives the fundamental theorem:

$$A'(x) = f(x).$$

The students found it quite natural to examine the graph stretched in this way and we felt that they had a better understanding of the fundamental theorem than other students following a more standard textbook approach.

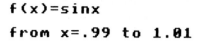

`f(x)=sinx`

`from x=.99 to 1.01`

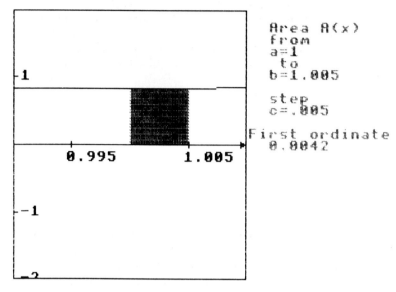

Figure 42. Computer picture of fundamental theorem A(x) = f(x)

What we did not do

Although students worked interactively with prepared software throughout the year we did not ask them to write their own programs or give detailed explanations of the construction of the programs in the package. This was a conscious decision, bearing in mind the scope of the present A-level syllabus and the restricted time available. But this is not to say that such developments would not have been desirable: the great majority of the students were interested in programming in BASIC, most having their own micros, and they often expressed interest in how the programs worked. It is easy to see that programs on the lines of '132 Short Programs for the Mathematics Classroom', published by the Mathematical Association, could be used in any A-level course, with the algorithms employed enhancing the understanding of many concepts. At present lack of time restricts such developments in the current A-level syllabus.

The upper sixth

Most of the foundations of the A-level syllabus have been laid in the first year and the computer played an integral part throughout. In the upper sixth the computer will continue to be available in every lesson. Our experience in using the computer at this level shows that it is most valuable when we meet such topics as the exponential function, logarithmic function and Taylor's series. However, it is likely to be used less than in the lower sixth as we become involved with answering standard examination questions. We still use the programs on occasion to check results graphically but, perhaps as important, the graphical images which students can now visualize in their mind's eye will be appealed to even in quick blackboard and chalk explanations.

The future
Our experience shows that the computer can be used as a useful adjunct to the current A-level syllabus. Even more mileage can be achieved through designing a new curriculum that takes the new possibilities into account. A fully integrated approach, with the students programming their own algorithms as well as using prepared software would require a provision for microcomputers currently beyond the financial resources of the average school. In this respect schools lag behind the provisions students have in their own homes. (Our class proved to have 15 computers among 16 students, only one not having his own machine.) One may look forward optimistically to a time when computers can realize a fuller potential in a much broader mathematics curriculum.

In the meantime it is clear that a single computer in a classroom can be used to advantage in the current syllabus. The same techniques should also prove helpful at an earlier age, in particular the exploratory investigations of a dynamic graphic approach on the computer can be used to give added meaning to the calculus concepts taught within any 16 + examination.

Note
The screen dumps used in this article were produced using 'Supergraph' and 'Graphic Calculus' published by Glentop Publishers, Standfast House, Bath Place, High Street, Barnet, Herts.